revise GCSE

Authors - Ian Honeysett, John Sadler and Carol Tear

Science

Contents

1 **Our bodies in action**

1.1	Food and digestion	09
1.2	Respiration	11
1.3	Response to stimuli	12
1.4	Homeostasis	17
1.5	Hormones and reproduction	20
How Science Works		23
Exam practice questions		25

2 **Health and disease**

2.1	Keeping healthy	27
2.2	Giving the body a helping hand	29
2.3	Drugs and health	32
2.4	Too much or too little	34
How Science Works		37
Exam practice questions		39

3 **Variation and genetics**

3.1	Genes and chromosomes	41
3.2	Manipulating genes	44
3.3	Variation and evolution	47
How Science Works		51
Exam practice questions		53

4 **Organisms and the environment**

4.1	The variety of life	55
4.2	Living together	57
4.3	Human impact on the environment	63
4.4	Conservation	66
How Science Works		68
Exam practice questions		70

5 **The building blocks**

5.1	Atoms	72
5.2	Bonding	74
5.3	Mixtures and compounds	76
5.4	Equations	81
5.5	The periodic table	82
How Science Works		87
Exam practice questions		89

6 **Non-metal chemistry**

6.1	Non-metals	91
6.2	Limestone, cement and concrete	93
6.3	Crude oil	94
6.4	Organic compounds of carbon	95
6.5	Polymers	98
6.6	The air	102
6.7	Nitrogen and its compounds	104
How Science Works		107
Exam practice questions		109

7 **The Earth and mineral chemistry**

7.1	The structure of the Earth	112
7.2	Rocks as building materials	114
7.3	Metals from rocks	116
7.4	Alloys	121
7.5	Salts	123
How Science Works		126
Exam practice questions		127

8 Reactions

8.1	Types of chemical reactions	131
8.2	Identification of gases	132
8.3	Rates of reaction	133
8.4	Energy transfer in reactions	136
	How Science Works	140
	Exam practice questions	142

9 Radioactivity

9.1	Radioactive emissions	145
9.2	Dangers of radiation	146
9.3	Uses of radioactivity	147
9.4	Changes in the nucleus	149
9.5	Nuclear reactors	151
	How Science Works	153
	Exam practice questions	154

10 Energy

10.1	The effect of heating	156
10.2	Heat transfer	158
10.3	Energy resources	160
10.4	How the electricity supply works	165
10.5	Electrical resistance	169
	How Science Works	171
	Exam practice questions	172

11 Waves

11.1	Describing waves	176
11.2	Seismic waves	177
11.3	Sound and ultrasound	179
11.4	Electromagnetic waves	179
11.5	Wave properties	180
	How Science Works	183
	Exam practice questions	185

12 Electromagnetic radiation

12.1	Energy and intensity	186
12.2	Wave communications	188
12.3	Radio waves and microwaves	189
12.4	Infrared and light	192
12.5	Ultraviolet, X-rays and gamma rays	193
12.6	The atmosphere and global warming	195
	How Science Works	197
	Exam practice questions	199

13 The Earth and beyond

13.1	The Earth	201
13.2	The Solar System	204
13.3	Gravity and orbits	206
13.4	Beyond the Solar System	207
13.5	Observing and exploring	209
13.6	The expanding Universe	212
	How Science Works	214
	Exam practice questions	216

Answers	217
Index	222

This book and your GCSE course

	AQA A	AQA B	EDEXCEL 360
Web address	www.aqa.org.uk		www.edexcel.org.uk
Specification number	4461	4462	2101
Modular tests	6 x 30 min objective tests. Two B, two C, two P 75%	3 x 45 min structured question tests. One B, one C, one P 75%	6 x 20 min objective tests. Two B, two C, two P 60%
Terminal papers	none	none	none
Availability of exams	Nov, March, June	Jan, June	Nov, March, June
Coursework	25%	25%	40%
BIOLOGY			
Our bodies in action	B1.11.1		B1b3
Health and disease	B1.11.2 B1.11.3 B1.11.4		B1b4
Variation and genetics	B1.11.6 B1.11.7		B1a1, B1a2
Organisms and the environment	B1.11.5 B1.11.8		B1a1
CHEMISTRY			
The building blocks	C1.12.1, C1.12.2		C1a5, C1b7, C1b8
Non-metal chemistry	C1.12.1, C1.12.3, C1.12.4, C1.12.6		C1a6, C1b7, C1b8, C1d6
The Earth and metal chemistry	C1.12.1, C1.12.2, C1.12.6		C1a6
Reactions	C1.12.1		C1a6
PHYSICS			
Radioactivity	P1.13.4, P1.13.5, P1.13.6		P1b11
Energy	P1.13.1, P1.13.2, P1.13.3, P1.13.4		P1a9, P1a10
Waves	P1.13.5		P1b11
Electromagnetic radiation	P1.13.5		P1b11
The Earth and beyond	P1.13.7		P1b12

Visit your awarding body for full details of your course or download your complete GCSE specifications.

Use these pages to get to know your course
- Make sure you know your exam board
- Check which specification you are doing

- Know how your course is assessed:
  – what format are the papers?
  – how is coursework assessed?
  – how many papers?

OCR A	OCR B	
www.ocr.org.uk		
J630	J640	
3 x 40 min objective style tests. Each paper contains B, C, P 50%	2 x 60 min structured tests. Each paper contains B, C, P 66.7%	
Ideas in context 16.7%	none	
Jan, June	Jan, June	
13.3%	33.3%	
B3.3	B1a, B1b, B1d, B1f	
B2.1, B2.2, B2.3, B2.4	B1a, B1b, B1c, B1e	
B1.1, B1.2, B1.3, B1.4, B3.1	B1g, B1h, B2f	
B3.4	B1b, B1c, B2a, B2b, B2c, B2d B2e, B2g, B2h	
C1.2, C2.2	C1a, C1b, C1c, C1d, C2d	
P2.4, C1.1, C2.2, C3.1	B2a, C1a, C1c, C1d, C1e, C1g, C2b, C2f, C2h	
P1.1, C1.2, C2.3	C1f, C2c, C2d, C2e	
C1.1, C2.3	C1h, C2g	
P2.1, P3.1, P3.2, P3.3, P3.4	P2c, P2d	
P3.3	P1a, P1b, P1c	
	P1d, P1e, P1f, P1g, P1h	
P2.1, P2.2, P2.3, P2.4, P3.2	P1d, P1e, P1f, P1h	
P1.1, P1.2, P1.3	P2e, P2f, P2g, P2h	

Preparing for the examination

Planning your study

The last few months before taking your final GCSE examinations are very important in achieving your best grade. However, the success can be assisted by an organised approach throughout the course. This is particularly important from 2007 as all the science courses are available in units.

- After completing a topic in school or college, go through the topic again in your Revise GCSE Science Study Guide. Copy out the main points on a sheet of paper or use a highlighter pen to emphasise them.
- Much of memory is visual. Make sure your notes are laid out attractively using spaces and symbols. If they are easy to read and attractive to the eye, they will be easier to remember.
- A couple of days later, try to write out these key points from memory. Check differences between what you wrote originally and what you wrote later.
- If you have written your notes on a piece of paper, keep this for revision later.
- Try some questions in the book and check your answers.
- Decide whether you have fully mastered the topic and write down any weaknesses you think you have.

Preparing a revision programme

Before an external examination, look at the list of topics in your examination board's specification. Go through and identify which topics you feel you need to concentrate on. It is a temptation at this time to spend valuable revision time on the things you already know and can do. It makes you feel good but does not move you forward.

When you feel you have mastered all the topics, spend time trying sample questions that can be found on your examination board's website. Each time, check your answers with the answers given. In the final week, go back to your summary sheets (or highlighting in the book).

How this book will help you

Revise GCSE Science Study Guide will help you because:

- it contains the essential content for your GCSE course without the extra material that will not be examined
- it contains GCSE Exam Practice Questions to help you to confirm your understanding
- examination questions from 2007 are different from those in the past. Trying past questions will not help you when answering some parts of the questions in 2007. The questions in this book have been written by experienced examiners.
- the summary table will give you a quick reference to the requirements for your examination

Four ways to improve your grade

1. Read the question carefully

Many students fail to answer the actual question set. Perhaps they misread the question or answer a similar question they have seen before. Read the question once right through and then again more slowly. Some students underline or highlight key words in the question as they read it through. Questions at GCSE contain a lot of information. You should be concerned if you are not using the information in your answer.

2. Give enough detail

If a part of a question is worth three marks you should make at least three separate points. Be careful that you do not make the same point three times. Draw diagrams with a ruler and label with straight lines.

3. Correct use of scientific language

There is important scientific vocabulary you should use. Try to use the correct scientific terms in your answers. The way scientific language is used is often a difference between successful and unsuccessful students. As you revise, make a list of scientific terms you meet and check that you understand the meaning of these words. Learn all the definitions. These are easy marks and they reward effort and good preparation.

4. Show your working

All science papers include calculations. Learn a set method for solving a calculation and use that method. You should always show your working in full. Then, if you make an arithmetical mistake, you may still receive marks for correct science. Check that your answer is given to the correct number of significant figures and give the correct units.

How Science Works

From 2007, all GCSE science courses must cover certain factual detail, similar to the detail that has been required for many years. Now, however, each course must also include study of 'How Science Works'.

This includes four main areas:

- **Data, evidence, theories and explanations**
  This involves learning about how scientists work, the differences between data and theories and how scientists form theories.

- **Practical skills**
  How to test a scientific idea including collecting the data and deciding how reliable and valid it is.

- **Communication skills**
  Learn how to present information in graphs and tables and to be able to analyse information that has been provided in different forms.

- **Applications and implications of science**
  Learning about how new scientific discoveries become accepted and some of the benefits, drawbacks and risks of new developments.

The different examining bodies have included material about how science works in different parts of their examinations. Often it is in the coursework but you are also likely to come across some questions in your written papers. Do not panic about this and think that you have not learnt this work. Remember these questions test your skills and not your memory; that is why the situations are likely to be unfamiliar. The examiners want you to show what you know, understand and can do.

To help you with this, there are sections at the end of each chapter called **How Science Works** and questions about how science works in the **Exam Practice Questions**. This should give you an idea of what to expect.

1 Our bodies in action

The following topics are covered in this chapter:

- ● Food and digestion
- ● Respiration
- ● Response to stimuli
- ● Homeostasis
- ● Hormones and reproduction

1.1 Food and digestion

What is in a balanced diet?

OCR B   B1b

All organisms require **food** to survive. It provides energy and the raw materials for growth. We take our food in ready-made as complicated organic molecules. These food molecules can be placed into seven main groups.

A **balanced diet** needs the correct amounts of each of these types of food molecules.

> Remember a balanced diet is the correct amount of each food, not simply 'enough'.

Food type	Made up of	Use in the body
carbohydrates	simple sugars, e.g. glucose	supply or store of energy
fats	fatty acids and glycerol	rich store of energy
proteins	long chains of amino acids	growth and repair
minerals	different elements, e.g. iron	iron is used to make haemoglobin
vitamins	different structures, e.g. vitamin C	prevents scurvy
fibre	cellulose	prevents constipation
water	water	all chemical reactions take place in water

The exact amount of each substance that is needed in a balanced diet will vary. It depends on how **old** the person is, whether they are **male or female** and how **active** they are.

For example, teenagers need a high-protein diet to provide the raw materials for growth. You can estimate the recommended daily average (RDA) protein intake for a person using the formula:

> There are differences between the sexes because of the time of the growth spurt and due to periods in girls.

$$\text{RDA in g} = 0.75 \times \text{body mass in kg}$$

However, it is not only the amount of protein that is important but also the type. Proteins from animals are called **first class proteins** because they contain more variety of **amino acids** compared with plant proteins.

Some people's diet may be influenced by other factors than just their daily requirements. Some people may be vegetarians or vegans and some religions require certain diets to be followed. Some people may have to avoid certain foods to prevent them becoming ill.

Digestion and absorption

OCR B   B1b

> **KEY POINT** The job of the digestive system is to break down large food molecules. This is called digestion.

Digestion happens in two main ways: **physical** and **chemical**. Physical digestion occurs in the mouth where the teeth break up the food into smaller pieces.

Chemical digestion is caused by digestive enzymes that are released at various points along the digestive system.

> Remember enzymes are biological catalysts found in all cells of the body.

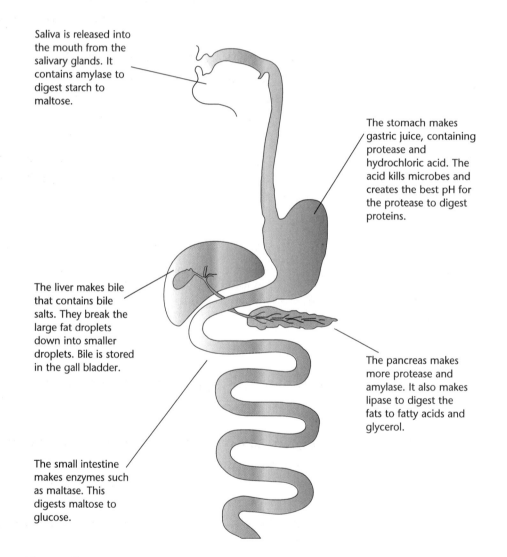

Saliva is released into the mouth from the salivary glands. It contains amylase to digest starch to maltose.

The stomach makes gastric juice, containing protease and hydrochloric acid. The acid kills microbes and creates the best pH for the protease to digest proteins.

The liver makes bile that contains bile salts. They break the large fat droplets down into smaller droplets. Bile is stored in the gall bladder.

The pancreas makes more protease and amylase. It also makes lipase to digest the fats to fatty acids and glycerol.

The small intestine makes enzymes such as maltase. This digests maltose to glucose.

Fig. 1.1 The digestive system.

Once the food molecules have been digested, they are small enough to diffuse into the bloodstream or lymph vessels. This is called **absorption**.

1.2 Respiration

Aerobic and anaerobic respiration

OCR B   B1b

Aerobic respiration

> **KEY POINT**
> Aerobic respiration is when glucose reacts with oxygen to release energy. Carbon dioxide and water are released as waste products.

$$\text{glucose} + \text{oxygen} \rightarrow \text{carbon dioxide} + \text{water} + \textbf{energy}$$
$$C_6H_{12}O_6 + 6O_2 \rightarrow 6CO_2 + 6H_2O + \text{energy}$$

We use the energy released from respiration for many processes. Respiration also gives off heat, which is used to maintain our high body temperature. Our rate of respiration can be estimated by measuring how much oxygen we use.
During exercise, the body needs more energy and so the rate of respiration increases. The breathing rate increases to obtain extra oxygen and remove carbon dioxide from the body. The heart beats faster so that the blood can transport the oxygen and carbon dioxide faster. This is why our pulse rate increases.

> It is actually the build up of carbon dioxide that makes us breathe faster.

Anaerobic respiration

> **KEY POINT**
> When not enough oxygen is available, glucose can be broken down by anaerobic respiration. This may happen during hard exercise.

In humans: $\text{glucose} \rightarrow \text{lactic acid} + \textbf{energy}$

Being able to respire without oxygen sounds a great idea. However, there are two problems:
● Anaerobic respiration releases less than half the energy of that released by aerobic respiration.
● Anaerobic respiration produces lactic acid. Lactic acid causes muscle fatigue and pain.
The build up of lactic acid is called the **oxygen debt**. After the exercise is finished, extra oxygen is needed by the liver to remove the lactic acid.

1.3 Response to stimuli

Patterns of response

All living organisms need to respond to changes in the environment.

Although this happens in different ways the pattern of events is always the same:

stimulus → detection → co-ordination → response

> Plants can also respond to stimuli but the response is usually slower than that of animals.

Detecting the stimulus

 KEY POINT   Receptors are specialised cells that detect a stimulus. Their job is to convert the stimulus into electrical signals in nerve cells.

Some receptors can detect several different stimuli but they are usually specialised to detect one type of stimulus:

Stimulus	Type of receptor
light	photoreceptors in the eye
sound	vibration receptors in the ears
touch, pressure, pain and temperature	different receptors in the skin
taste and smell	chemical receptors in the tongue and nose
position of the body	receptors in the ears

A **sense organ** is a group of receptors gathered together with some other structures.

The other structures help the receptors to work more efficiently. An example of this is the eye.

Co-ordination

The body receives information from many different receptors at the same time.

 KEY POINT   Co-ordination involves processing all the information from receptors so that the body can produce a response that will benefit the whole organism.

In most animals this job is done by the central nervous system (CNS).

Response

KEY POINT   Effectors are organs in the body that bring about a response to the stimulus.

Usually these effectors are muscles and they respond by contracting. They could however be glands and they may respond by releasing an enzyme. Many responses are **reflexes**.

An example of a sense organ – the eye

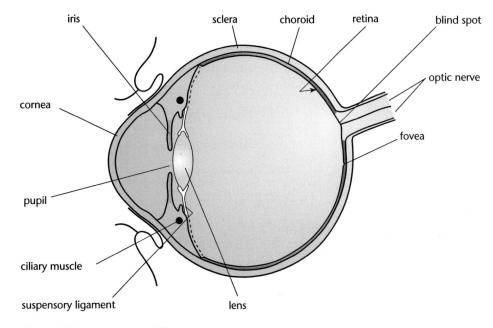

Fig. 1.2 The structure of the eye.

The light enters the eye through the pupil. It is focused onto the **retina** by the **cornea** and the **lens**. The size of the **pupil** can be changed by the muscles of the **iris** when the brightness of the light changes. The aim is to make sure that the same amount of light enters the eye. The job of the lens is to change shape so that the image is always focused on the light-sensitive retina.

The receptors are cells in the retina called rods and cones. They detect light and send messages to the brain along the **optic nerve**.

The lens must be a different shape when the eye looks at a close object compared with a distant object. This is to make sure that the light is always focused on the back of the retina. The ciliary muscle changes the shape of the lens as shown in the diagram. This is called **accommodation**.

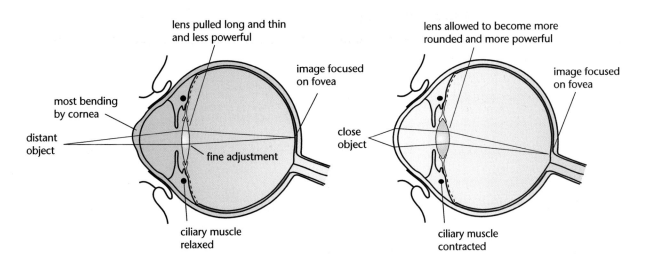

Fig. 1.3 How the eye focuses.

Some people have problems with their eyes:

Condition	Cause	Treatment
long or short sight	the eyeball or lens is the wrong shape	long sight and short sight can be corrected by wearing convex or concave lenses respectively; cornea surgery can now also be used
red–green colour blindness	lack of certain cones in the retina	no treatment
poor accommodation	lens becomes less elastic in senior citizens	wearing glasses with half convex and half concave lenses

The eyes are also used to judge **distances**. Animals that hunt usually have their eyes on the front of their head. Each eye has a slightly different image of the object. This is called binocular vision and it can be used to judge distance. Animals that are hunted usually have eyes on the side of their heads. This gives monocular vision and they cannot judge distances so well. They can, however, see almost all around.

Neurones and synapses

To communicate between receptors and effectors the body uses two **main** methods.

These are:

- **nerves**
- **hormones**.

> **KEY POINT**
>
> A neurone is a specialised cell that is adapted to pass electrical impulses.

The **central nervous system** (CNS) contains millions of neurones but outside the CNS, neurones are grouped together into bundles of hundreds or thousands. These bundles are called nerves.

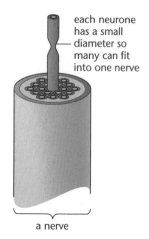

each neurone has a small diameter so many can fit into one nerve

a nerve

Fig. 1.4 The structure of a nerve.

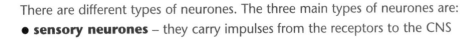

There are different types of neurones. The three main types of neurones are:
- **sensory neurones** – they carry impulses from the receptors to the CNS

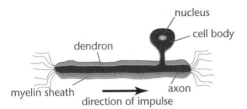

Fig. 1.5 Sensory neurone.

- **motor neurones** – they carry impulses from the CNS to the effectors

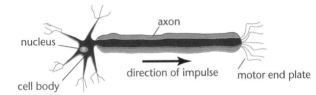

Fig. 1.6 Motor neurone.

- **relay neurones** – they pass messages between neurones in the CNS.

Although all neurones have different shapes, they all have certain features in common:
- One or more long **projections** from the cell body to carry the impulse a long distance.
- A fatty covering (**myelin sheath**) around the projection for insulation.
- Many fine endings (**dendrites**) so that the impulse can be passed on to many cells.

> Exam questions often ask how neurones are adapted for their job.

Synapses

Each neurone does not directly end on another neurone.

There is a small gap between the two neurones called a **synapse**.

In order for an impulse to be generated in the next neurone, a **chemical transmitter** is released. This then diffuses across the small gap.

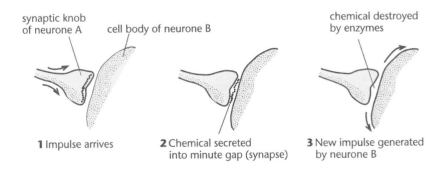

Fig. 1.7 Chemical transmission between nerves.

Many drugs work by interfering with synapses. They may block or copy the action of neurotransmitters in certain neurones.

Types of response

The nervous system is made up of the CNS and the **peripheral nervous system**.

 KEY POINT The CNS is the brain and spinal cord. The peripheral nervous system is all the nerves passing information to and from the CNS.

Once the information reaches the CNS from a sensory neurone there is a choice:
Either: The message may be passed straight to a motor neurone via a relay neurone. This is very quick and is called a **reflex action**.

Or: The message can be sent to the higher centres of the brain and the organism might decide to make a response. This is called a **voluntary action**.

A reflex action

All reflexes are:

* fast
* do not need conscious thought
* protect the body.

Do not say that reflexes do not involve the brain. Reflexes such as the pupil reflex go via the brain.

Examples of reflexes include the knee jerk, pupil reflex, accommodation, ducking and withdrawing the hand from a hot object.

This diagram shows the pathway for a reflex that involves the spinal cord:

1 Stimulus is detected by sensory cell.

↓

2 Impulse passes down sensory neurone.

↓

3 Relay neurone passes impulse to motor neurone.

↓

4 Motor neurone passes impulse to effector.

↓

5 Muscle contracts.

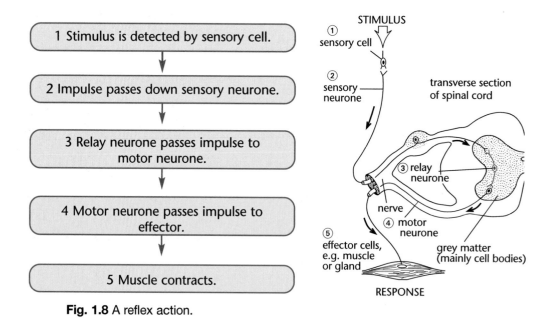

Fig. 1.8 A reflex action.

A voluntary action

Voluntary actions need a conscious decision in order to take place. They therefore always involve the **brain**.

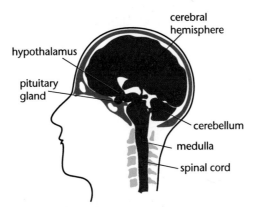

Fig. 1.9 The structure of the brain.

The **cerebral hemisphere** is the area of the brain where the decisions are made. Nerve impulses from here are sent down the spinal cord to effectors via motor neurones.

1.4 Homeostasis

Homeostasis and hormones

> **KEY POINT**
> It is vital that the internal environment of the body is kept fairly constant. This is called homeostasis.

The different factors that need to be kept constant include:

water content   temperature   sugar levels   mineral content

Many of the mechanisms that are used for homeostasis involve hormones.

> **KEY POINT**
> Hormones are chemical messengers that are carried in the blood stream.

They are released by glands and pass to their target organ.

Only Edexcel candidates need to know the structure of the brain.

Hormones are carried dissolved in the plasma of the blood.

Figure 1.10 shows the main hormone-producing glands of the body. Between them they make a number of different hormones:

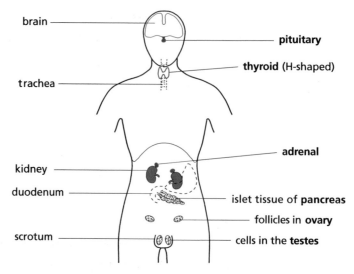

Fig. 1.10 Hormone-producing glands.

Hormones take longer to have an effect than nerves but their responses usually last longer.

Many of these control mechanisms work by **negative feedback**. This means that if the levels change too much, a hormone is released and this brings the change back to the normal level.

Control of blood sugar

OCR B B1f
EDEXCEL 360 B1b3

It is vital that the sugar or glucose level of the blood is kept constant. If it gets too low then cells will not have enough to use for respiration. If it is too high then glucose may start to pass out in the urine.

KEY POINT   Insulin is the hormone that controls the level of glucose in the blood.

When glucose levels are too high, more insulin is made. The insulin converts excess glucose into glycogen to be stored in the liver:

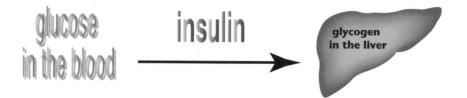

People with **diabetes** do not produce enough insulin naturally. They need regular insulin injections in order to control the level of glucose in their blood. They also need to control their diet carefully.

Control of body temperature

OCR B   B1f

It is important to keep our body temperature at about 37 °C. This is because it is the best temperature for enzymes to work.

The blood temperature is monitored by the brain and if it varies from 37 °C, various changes are brought about.

> **Camels can put up with their body temperature rising by 6°C without their enzymes being damaged.**

When we feel too hot

When we feel hot we need to lose heat faster, as our core body temperature is in danger of rising.

We do this by:
- **sweating** – as water evaporates from our skin, it absorbs heat energy. This cools the skin and the body loses heat.
- **vasodilation** – blood capillaries near the skin surface get wider to allow more blood to flow near the surface. Because the blood is warmer than the air, it cools down and the body loses more heat.

If the blood temperature gets too high it could lead to heat stroke and dehydration.

When we feel too cold

When we feel too cold we are in danger of losing heat too quickly and cooling down. This means we need to conserve our heat to maintain a constant 37 °C.

We do this by:
- **shivering** – rapid contraction and relaxation of body muscles. This increases the rate of respiration and more energy is released as heat
- **vasoconstriction** – blood capillaries near the skin surface get narrower and this process reduces blood flow to the surface. The blood is diverted to deeper within the body to conserve heat.
- **sweating less**.

Hypothermia occurs when the blood temperature gets too low. It can be fatal.

> **If vasoconstriction occurs for a long time it can lead to frostbite.**

1.5 Hormones and reproduction

The reproductive hormones

OCR B   B1f
EDEXCEL 360   B1b3

Hormones are responsible for controlling many parts of the reproduction process.

This includes:

- the development of the sex organs
- the production of sex cells
- controlling pregnancy and birth.

The main hormones controlling these processes are shown in the table.

Hormone	Male or female	Produced by the	Main function
testosterone	male	testes	stimulates the male secondary sexual characteristics
oestrogen	female	ovaries	stimulates the female secondary sexual characteristics; repair of the wall of the uterus; controls ovulation
progesterone	female	ovaries and placenta	prevents the wall of the uterus breaking down.

Testosterone and oestrogen control the changes occurring in the male and female bodies at puberty. These changes are the **secondary sexual characteristics**:

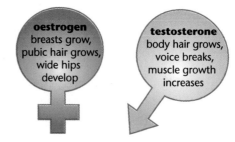

oestrogen
breasts grow, pubic hair grows, wide hips develop

testosterone
body hair grows, voice breaks, muscle growth increases

The secondary sexual characteristics also include the production of the sex cells. In the male they are **sperm** and in females they are **eggs**.

After puberty, sperm are produced continuously but in the female one egg is usually released about once a month.

This means that oestrogen and progesterone levels vary at different times in the monthly cycle.

Oestrogen levels are high in the first half of the cycle. The oestrogen prepares the wall of the uterus to receive a fertilised egg. It does this by making the wall thicker and increasing its supply of blood. It also triggers the release of an egg. This is called **ovulation**.

Progesterone is high in the second half of the cycle. It further repairs the wall of the uterus and stops it breaking down.

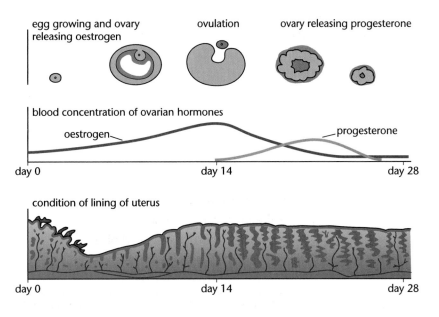

Fig. 1.11 Changes occurring during the monthly cycle.

Using hormones to control reproduction

OCR B   B1f
AQA   B1.11.1
EDEXCEL 360   B1b 3

The production of oestrogen and progesterone is controlled by the release of other hormones. These hormones are made in the **pituitary gland** in the brain.

> **OCR A and Edexcel candidates need to know how hormones can control reproduction but do not need to know the roles of FSH and LH.**

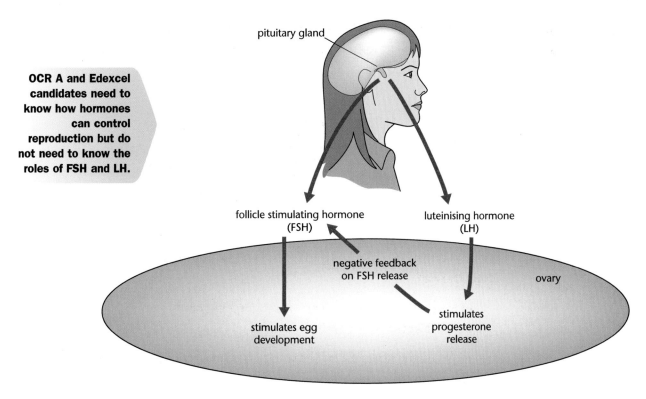

Fig. 1.12 Control of reproduction.

It is now possible to produce synthetic versions of these hormones. They can be used to control the fertility of women.

This can happen in two main ways.

Increasing fertility

Some women find it difficult to get pregnant because they do not produce eggs regularly. These women can take a **fertility drug**. This contains hormones that are similar to FSH. The drugs stimulate the production of eggs and sometimes a number of eggs are released each month.

Sometimes women are treated with fertility drugs and the eggs are removed from their body. The eggs can then be fertilised by sperm outside the body. The embryo can then be put back inside the uterus. This process is called *in vitro* fertilisation or IVF. It can be used on women who have blocked oviducts.

Decreasing fertility

Scientists are developing a male contraceptive pill that would stop sperm production.

Other women may want to stop themselves becoming pregnant. They take drugs that are called **oral contraceptives**. These drugs contain hormones that prevent the pituitary gland releasing FSH. This means that the ovary will not produce eggs.

HOW SCIENCE WORKS

Parkinson's disease and smoking – is it worth the risk?

Parkinson's disease is a disorder of the central nervous system that is caused by a loss of cells in an area of the brain.

Those cells produce dopamine, a chemical messenger responsible for transmitting signals across synapses. Loss of dopamine causes some nerve cells to fire out of control, leaving patients unable to control their movement in a normal way.

Patients often start off with slight shaking of the hands and they may eventually have difficulty walking, talking or completing other simple tasks.

The disease can be treated by giving a drug called L-dopa. The brain uses this chemical to make more dopamine. Unfortunately, the drug has a number of side effects and larger doses of the drug are often needed as the disease gets worse. Patients therefore have a difficult decision to make.

In recent studies on Parkinson's disease it has been suggested that there is a **correlation** between smoking and a reduced risk of having the disease.

Doctors are trying to find out if this means that smoking actually **causes** the protection.

Do these results mean that doctors are actually advising patients to smoke? No, they are not.

As a doctor said:

❝The dangers of cigarette smoking far outweigh any as yet inconclusive evidence that there are advantages of protection from Parkinson's disease.

You need to look at the risks – smoking is the largest single cause of preventable death in many countries.

About 1 in 1000 people are likely to get Parkinson's disease but every year over 100 000 people die through cigarette smoking in the UK.

Smoking is just not worth the risk. ❞

HOW SCIENCE WORKS

Treating infertility – the risks and ethics

Infertility is a much greater problem than many people realise. The figures below give some details about infertility in the UK.

- About one in seven UK couples have difficulty having children – approximately 3.5 million people.
- Though many of these will become pregnant naturally given time, quite a few will not.
- Infertility is the most common reason why women aged 20–45 go to see their doctor, after pregnancy itself.

There is no doubt that infertility is a major problem in the UK but views on what to do about it vary.

Trying to decide who should have fertility treatment is very difficult. There have been a number of uses of fertility treatment that have worried people.

- Some people think that infertile couples should be allowed one treatment paid for by the National Health Service. The trouble is that the chance of success after the first treatment is about 10%. This increases with each treatment.
- Taking fertility drugs may cause a woman to become pregnant with a large number of babies. The chance of some of these babies surviving can be increased if some of the embryos are removed at an early stage.
- Women who are too old to have children can be treated with hormones so that they can get pregnant and give birth.

Doctors have to make decisions about who to treat all the time. Sometimes these decisions involve balancing **risks** and sometimes they involve **ethical** judgements.

❝I think that fertility treatment by drugs or IVF treatment is too expensive and should not be paid for by the National Health Service. ❞

❝I think that people who are infertile have a right to have treatment. Being unable to have children can make them depressed and this is as serious as other illnesses. ❞

Exam practice questions

1. Which hormone in the list causes a boy's voice to deepen at puberty?
   (a) insulin
   (b) oestrogen
   (c) progesterone
   (d) testosterone   [1]

2. Which of the following are female secondary sexual characteristics?
   (a) growth of breasts
   (b) sperm production
   (c) presence of ovaries
   (d) presence of a uterus   [1]

3. Which of the following changes might happen in our body when we get too hot?
   (a) sweating increases
   (b) blood flow to the skin drops
   (c) we shiver more
   (d) blood vessels in the skin close up   [1]

4. Finish the sentences using words from the list.

   A balanced diet contains seven groups of substances. Proteins are necessary for
   _____ and are made from molecules called _____ _____. Foods such as
   minerals are another group of substances. An example is _____, which is needed to
   make haemoglobin. Many molecules such as proteins are too large to be able to pass into
   our blood stream and so need to be _____ first. The process of taking food into the
   blood stream is called _____.   [5]

   **absorption   amino acids   digested   egested   growth   iron
   respired   storing energy   sugars   vitamin C**

5. The diagram shows some of the organs inside the body.

   (a) Write down the name of the organ shown that:
       (i) makes insulin
       (ii) stores glycogen   [2]
   (b) It is very important to keep the level of glucose
       in the blood roughly constant. Explain why.   [2]
   (c) A person has a drink of a liquid containing
       glucose.
       Explain the role of the organs shown in the
       diagram in dealing with this glucose.
       You should include:
       • how the glucose gets into the body
       • what happens when the blood glucose level rises
       • what happens to the glucose.   [5]

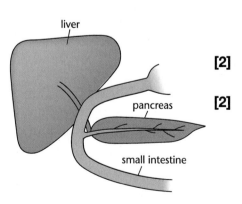

Exam practice questions

6. The graph shows the level of two hormones in a woman and the woman's body temperature at different days of her monthly cycle.

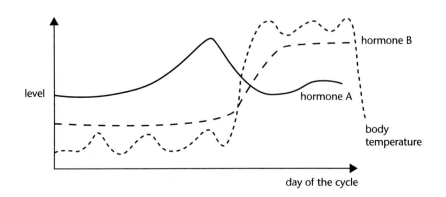

(a) Copy out the following table and complete the blank boxes.

Hormone	Name of hormone	In which half of the cycle does the hormone reach its highest level?
Hormone A		
Hormone B		

[4]

(b) Write down the name of the organ where hormones A and B are made. [1]

(c) Describe how the body temperature changes during the cycle. [2]

(d) Some women measure their body temperature very accurately. They then use this to prevent getting pregnant or to help them get pregnant.
(i) How could this information help them to get pregnant?
(ii) How could it help to prevent them getting pregnant? [3]

(e) Suggest why using body temperature measurements is not a good method on its own to prevent pregnancy. [1]

7. Look back at the How Science Works article about infertility.
Treating fertility can produce some difficult ethical issues.

(a) Explain the difficult choices facing a woman who has become pregnant with a large number of embryos. [2]

(b) Suggest why some people think that it is not a good idea for women aged over 60 to have children. [2]

(c) What are the advantages and disadvantages of allowing the NHS to pay for only one IVF treatment per couple? [2]

2 Health and disease

The following topics are covered in this chapter:

- **Keeping healthy**
- **Giving the body a helping hand**
- **Drugs and health**
- **Too much or too little**

2.1 Keeping healthy

Causes of disease

OCR B · B1a
AQA · B1.11.4

A disease occurs when the normal functioning of the body is disturbed. **Infectious diseases** can be passed on from one person to another but **non-infectious diseases** cannot.

Type of disease	Description	Examples
Non-infectious • body disorder	incorrect functioning of a particular organ	diabetes, cancer
• deficiency disease	lack of a mineral or vitamin	anaemia, scurvy
• genetic disease	caused by a defective gene	red–green colour blindness
Infectious disease	caused by a pathogen	TB

 Genetic diseases are covered on page 44.

KEY POINT — Organisms that cause infectious diseases are called pathogens.

A number of different types of organisms can be pathogens.

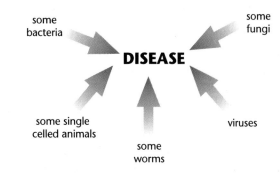

Fig. 2.1 Pathogenic organisms.

Pathogens may reproduce rapidly in the body, either damaging cells directly or producing chemicals called toxins, which make us feel ill.

Viruses damage cells by taking over the cell and reproducing inside them.

How the body protects itself

OCR A B2.1
OCR B B1c
AQA B1.11.4
EDEXCEL 360 B1b4

The skin covers most of the body and is quite good at stopping pathogens entering the body.

The body has a number of other defences that it uses in order to try to stop pathogens entering.

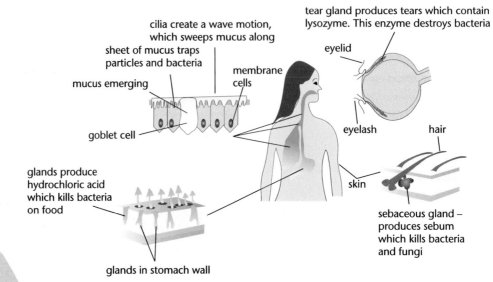

> **Acid in the stomach also provides the best pH for protease enzymes to work.**

Fig. 2.2 The body's defences.

If the pathogens do enter the body then the body will attack them in a number of ways.

The area that is infected will often become **inflamed** and two types of **white blood cells** attack the pathogen.

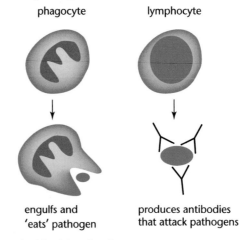

Fig. 2.3 The actions of white blood cells.

The pathogen is detected by the white blood cells because it has foreign chemical groups called **antigens** on its surface. The **antibodies** that are produced by the body are specific to a particular pathogen or toxin and will only destroy that particular antigen.

2.2 Giving the body a helping hand

Antibiotics and antiseptics

OCR A   B2.2/3
OCR B   B1c
AQA   B1.11.4
EDEXCEL 360   B1b4

Antibiotics

Sometimes a pathogen can make us ill before our body's immune system can destroy it. We may sometimes need to take drugs called **antibiotics** to kill the pathogen.

 KEY POINT   Antibiotics are chemicals that are usually produced by microorganisms that kill bacteria and fungi. They do not have any effect on viruses.

The first antibiotic to be widely used was **penicillin**. Today there are a number of different antibiotics that are used to treat different bacteria. This has meant that some diseases that once killed millions of people can now be treated.

There is a problem, however. More and more strains of bacteria are appearing that are **resistant to antibiotics**.

This is an example of natural selection which is covered on page 49.

A genetic change, or **mutation**, in the bacteria population can enable a large population of resistant bacteria to appear.

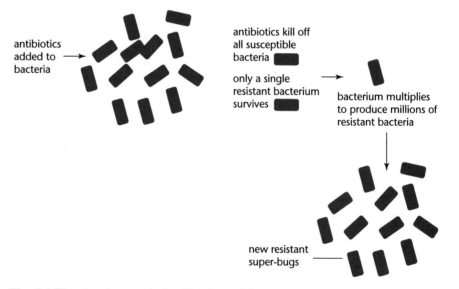

antibiotics added to bacteria

antibiotics kill off all susceptible bacteria

only a single resistant bacterium survives

bacterium multiplies to produce millions of resistant bacteria

new resistant super-bugs

Fig. 2.4 The development of antibiotic resistance.

This process has occurred in many different types of bacteria, including the TB causing bacterium and one called MRSA. These bacteria are now resistant to many different types of antibiotic and so are very difficult to treat.

There are various ways that doctors try to prevent the spread of these resistant bacteria:

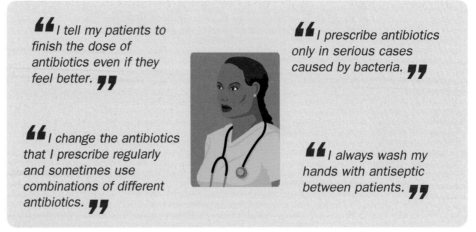

> 66 I tell my patients to finish the dose of antibiotics even if they feel better. 99

> 66 I prescribe antibiotics only in serious cases caused by bacteria. 99

> 66 I change the antibiotics that I prescribe regularly and sometimes use combinations of different antibiotics. 99

> 66 I always wash my hands with antiseptic between patients. 99

Antiseptics

One important weapon against resistant bacteria is the use of **antiseptics**.

> **KEY POINT**
> Antiseptics are artificial chemicals that kill pathogens outside the body.

> An antiseptic is usually used on the body and a disinfectant is usually used on other surfaces.

They were first used by an Austrian doctor called Dr Semmelweis to sterilise medical instruments. The use of antiseptics in hospitals is vital in preventing the spread of resistant bacteria.

Vaccinations

OCR A   B2.1/2
OCR B   B1c
AQA   B1.11.4
EDEXCEL 360   B1b4

How do vaccines work?

When our body encounters a pathogen, white blood cells make antibodies against the pathogen. If they encounter the same pathogen again in the future then antibodies are produced faster and the pathogen is killed more quickly. This is called **immunity**.

This idea has been used in **vaccinations**.

> An immunisation is another name for a vaccination.

> **KEY POINT**
> A vaccine contains harmless versions of the pathogen which stimulate immunity.

A vaccination stimulates our body to make antibodies and special cells called **memory cells**. When the real pathogen comes along, it will be destroyed quickly.

This type of immunity where the antibodies are made by the person is called **active immunity**.

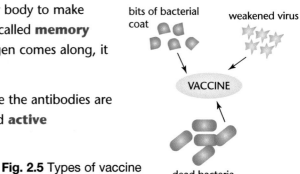

Fig. 2.5 Types of vaccine producing active immunity.

Sometimes it might be too late to give somebody this type of vaccination because they already have the pathogen. They can be given an injection containing antibodies made by another person or animal.

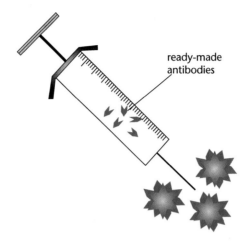

ready-made antibodies

Fig. 2.6 A vaccination containing antibodies.

A baby can also get antibodies from breast milk.

This is called **passive immunity**. A similar thing happens when a baby receives antibodies from its mother across the placenta.

How long do vaccines protect us for?

Some vaccines protect us for a long time because the memory cells survive for many years. The problem is that some diseases such as influenza need new vaccinations every year.

This is because the virus mutates and changes the shape of its outer coat. This means that different antibodies are needed, so a different vaccination is required.

The human immunodeficiency virus (HIV) also mutates regularly and weakens the immune system. This is making it very difficult to produce a vaccine.

For a vaccine to reduce or completely get rid of a disease, most of the population must be treated. This has sometimes proved difficult as some people are worried about the side effects of certain vaccines.

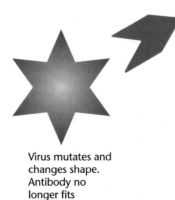

Virus mutates and changes shape. Antibody no longer fits

New antibody has to be made to fit the shape of the virus

Fig. 2.7 Producing a vaccine.

2.3 Drugs and health

Types of drugs

OCR B   B1e
AQA   B1.11.3
EDEXCEL 360   B1b4

KEY POINT

Drugs are chemicals that alter the functioning of the body.

Some drugs such as antibiotics are often beneficial to our body if used correctly. Others can be harmful, particularly those that are used recreationally.

> **Sedatives and stimulants often affect the action of synapses (see page 15.)**

Many drugs are **addictive**. This means that people want to carry on using them even though they may be having harmful effects. If they stop taking them they may suffer from unpleasant side effects called **withdrawal symptoms**. It also means that people develop **tolerance** to the drug, which means that they need to take bigger doses to have the same effect. Heroin and cocaine are very addictive.

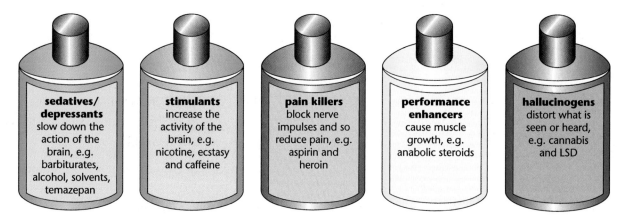

sedatives/depressants	stimulants	pain killers	performance enhancers	hallucinogens
slow down the action of the brain, e.g. barbiturates, alcohol, solvents, temazepan	increase the activity of the brain, e.g. nicotine, ecstasy and caffeine	block nerve impulses and so reduce pain, e.g. aspirin and heroin	cause muscle growth, e.g. anabolic steroids	distort what is seen or heard, e.g. cannabis and LSD

Fig. 2.8 Different drugs do different things.

In order to control drugs, many can only be bought with a prescription. Illegal drugs are classified into groups. Class A drugs are the most dangerous, and class C are the least dangerous. If people are caught with illegal class A drugs the penalties are the highest.

Smoking and drinking alcohol

OCR B   B1e
AQA   B1.11.3
EDEXCEL 360   B1b4

Smoking

Many people cannot give up smoking **tobacco** because it contains the drug **nicotine**. This is addictive. The nicotine is harmful to the body but most damage is done by the other chemicals in the tobacco smoke.

- Chemicals in the tar may cause cells in the lung to divide uncontrollably. This can cause **lung cancer**.

- The heat and chemicals in the smoke destroy the cilia on the cells lining the airways. The goblet cells also produce more mucus than normal. The bronchioles may become infected. This is called **bronchitis**.

- The mucus collects in the alveoli and may become infected. This may lead to the walls of the alveoli being damaged. This reduces gaseous exchange and is called **emphysema**.

- The nicotine can cause an increase in blood pressure increasing the chance of a **heart attack**.

Fig. 2.9 Problems resulting from smoking.

> Unlike oxygen, carbon monoxide does not let go of haemoglobin very easily.

Smoking tobacco is particularly dangerous for pregnant women. The **carbon monoxide** in the smoke combines with oxygen in the mother's blood and this can deprive the baby of oxygen. This may lead to a low birth weight.

Drinking alcohol

Drinking alcohol can have a number of effects on the body:

Short-term effects	Long-term effects
upsets balance and muscle control	damage to the liver (cirrhosis)
blurred vision and speech	brain damage
slower reactions	
helps people relax	

Owing to the effects of alcohol on the body there is a legal limit for the level of alcohol in the blood of drivers and pilots.

2.4 Too much or too little

Diseases of excess

OCR A   B2.4
OCR B   B1b
AQA   B1.11.2

It is important to maintain a balanced diet for the healthy functioning of the body. In the developed world many people eat too much food. This can make a person more likely to get various diseases.

Obesity

If a person eats food faster than it is used up by the body then the excess will be stored. Much of this will be stored as fat and can lead to **obesity**.

Obesity can be linked to a number of different health risks:

- **arthritis** – the joints wear out
- **diabetes** – unable to control the blood sugar level
- **breast cancer**
- **high blood pressure**
- **heart disease**.

> Scientists think that up to a million people in the UK might have diabetes now and many do not know they have it.

It is possible to estimate if a person is underweight, normal, overweight or obese by using the formula:

> Doctors take blood pressure using an inflatable cuff around the arm.

Body Mass Index (BMI) =

$$\frac{\text{mass in kg}}{(\text{height in metres})^2}$$

The BMI figure can then be checked in a table to see what range a person is in.

Blood pressure

Contractions of the heart pump blood out into the arteries under pressure. This is so it can reach all parts of the body.

Doctors often measure the **blood pressure** in the arteries and give two figures for example 120 over 80. The highest figure is called the **systolic pressure** and is the pressure when the heart contracts. The second figure is when the heart is relaxed and is called the **diastolic pressure**.

Blood pressure varies depending on various factors. The following factors can increase blood pressure:

- high salt and fat in the diet
- stress
- lack of exercise
- obesity
- high alcohol intake
- ageing.

A blood pressure that is too high or too low can cause problems in the body:

too low

dizziness
fainting
poor circulation

blood pressure

too high

burst blood
vessels
strokes
kidney damage

Heart disease

> Angina is a pain in the shoulders and arm caused by a lack of blood reaching the heart muscle.

The heart is made up of muscle cells that need to contract throughout life. This needs a steady supply of **energy** so the cells need **oxygen** and **glucose** at all times for **respiration**. This is supplied by blood vessels.

Fatty deposits can form in these blood vessels and reduce the flow of oxygen and glucose to the heart muscle cells.

> **KEY POINT** This reduction in blood flow causes heart disease and if an area of muscle stops beating then this is a heart attack.

There are many factors that can make it more likely for a person to have heart disease:

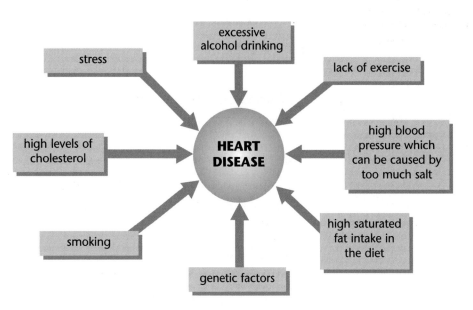

excessive alcohol drinking

stress

lack of exercise

high levels of cholesterol

HEART DISEASE

high blood pressure which can be caused by too much salt

smoking

high saturated fat intake in the diet

genetic factors

Cholesterol is an important chemical in the body. It is made in the liver and carried in the blood. In the blood it is carried by chemicals called lipoproteins. There are two types of these, called **low-density lipoproteins** (LDLs) and **high-density lipoproteins** (HDLs). The balance of these is important as LDLs can increase the chance of having heart disease and HDLs can protect the heart.

Saturated fats increase LDLs and so are bad for you, but polyunsaturated fats increase HDLs.

Deficiency diseases

In the developing world many people cannot get a balanced diet because there is not enough food to eat.

> **KEY POINT**
>
> Eating too little of one type of food substance can lead to a deficiency disease.

Examples of **deficiency diseases** are:
- **anaemia** due to a lack of iron
- **scurvy** due to a lack of vitamin C
- **kwashiorkor** due to a lack of protein.

There are times when people do not eat enough food although there is food available. They may put themselves on a diet because they have a poor self-image or think that they are overweight when they are not. This can reduce their resistance to infection and cause irregular periods in women. It may lead to illnesses such as **anorexia**.

HOW SCIENCE WORKS

Drug testing – an ethical problem

New drug makes men dangerously ill

In March 2006, six men volunteered to take part in tests on a new drug. The drug had been designed to treat serious diseases such as rheumatoid arthritis, leukaemia and multiple sclerosis.

This was the first time the drug had been used on people and unfortunately it caused serious side effects on the men, who became very ill. But why was this drug tested on people and had it been tested elsewhere first?

- Once a new drug has been made, it is tested on cells in a laboratory.
- If it passes these tests it is then tried on animals.
- Then it is tested on human volunteers.
- Finally, it is tested on human patients with the disease.

Many of these tests cause disagreements.

Many people think that animals should not be used to test drugs. Some think that it is too cruel, others think that it is pointless.

"The case in March 2006 shows that it is pointless testing drugs on animals. They do not react the same way as people."

"We must carry on testing on animals. This has prevented thousands of harmful drugs being tested on people."

Once the drug is cleared to be tested on patients the trial has to be set up carefully. One group are given the drug and another group have a **placebo**. This looks like the real treatment but has no drug in it.

The two groups do not know which treatment they are having, nor does the doctor who treats them. This is called a **double blind test** and it means that the people are not influenced by knowing which treatment is being given.

There are many ethical issues involved in testing drugs.

Some people think that tests are cruel or dangerous and therefore should not be done.

Others think that the tests are reasonable because the **benefits outweigh the disadvantages** of the tests.

HOW SCIENCE WORKS

The risks of vaccinations

Whether or not to give your children vaccinations is a difficult decision for some people to make. Diseases such as measles, mumps and rubella can have serious effects on the body.

- **Measles** is a very serious disease. 1 in 2500 babies that catch the disease die.
- **Mumps** may cause deafness in young children. It may also cause viral meningitis which can be fatal.
- **Rubella** can cause a baby to have brain damage if its mother catches the disease during pregnancy.

The graph shows the number of people getting measles in one country (Ireland) each year.

A measles vaccine was introduced in 1985.

In 1988, a combined measles, mumps and rubella (MMR) vaccine was introduced.

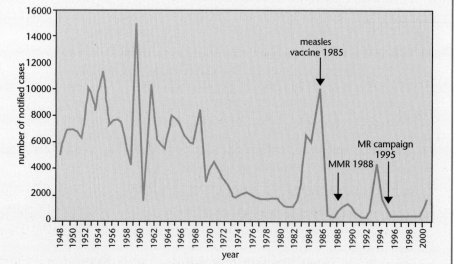

A measles and rubella (MR) vaccination campaign for primary school-age children was conducted in 1995.

The introduction of measles vaccine and the MMR vaccine has led to a decrease in measles. However the uptake of MMR was not high enough to prevent outbreaks in 1993 and 2000.

Over 1600 cases of measles were notified in Ireland in 2000. Measles has been eliminated or is close to elimination in Finland, Spain and other European countries where there is good uptake of vaccines. In Ireland, eight deaths from measles were reported between 1990 and 1999 and in 2000 three children in Dublin died from the disease.

So why are people worried about the vaccination?

In 1998, a study of autistic children raised the question of a connection between MMR vaccine and autism. (People with autism have difficulty with communicating and using some thinking skills.)

The 1998 study has a number of limitations. For example, the study was very small, involving only 12 children. This is too small a sample to make any generalisations about the causes of autism. In addition, the researchers suggested that MMR vaccination caused bowel problems in the children, which then led to autism. However, in some of the children studied, symptoms of autism appeared *before* symptoms of bowel disease.

Some people say that children should be forced to have the vaccine otherwise the disease will not disappear.

It is impossible to say that having a vaccine does not involve a **risk**. However, like many things in life it is a question of balancing risks, the risks of the vaccination against the risk of the disease.

Exam practice questions

1. Which part of the blood produces antibodies?
   (a) red blood cells
   (b) white blood cells
   (c) platelets
   (d) plasma   **[1]**

2. Which of the following could increase the risk of getting heart disease?
   (a) more stress
   (b) less saturated fat in the diet
   (c) stopping smoking
   (d) drinking a glass of wine a day   **[1]**

3. Lack of vitamin C in the diet could cause
   (a) kwashiorkor
   (b) anorexia
   (c) anaemia
   (d) scurvy   **[1]**

4. Fill in the gaps in the following sentences using words from the list:
   An organism that causes a disease is called a _____. The body tries to
   prevent these organisms entering the body. Tears contain _____ and the
   stomach makes _____, both of which can destroy the organisms.

   If the organism enters the body it can damage cells or release poisons called

   _____.   **[4]**

   acid   antibodies   lysozyme   pathogen   sebum   toxins

5. The MMR vaccination is usually given to young children. It protects them from measles,
   mumps and rubella. Here is some information about these three diseases.

   ● **Measles** is a very serious disease. 1 in 2500 babies that catch the disease die.
   ● **Mumps** can cause deafness in young children.
   ● **Rubella** can cause a baby to have brain damage if its mother catches the disease during
   pregnancy.

   (a) The MMR vaccination is given to young children rather than waiting until they
       are older. Write down **one** reason why.   **[1]**
   (b) Pregnant women are tested to see if they have had the MMR vaccination.
       Why is this important?   **[1]**
   (c) Explain how a vaccination such as MMR can protect a person from getting a
       disease.   **[3]**
   (d) People often feel ill after having a vaccination like MMR. Explain why this is. **[2]**

Exam practice questions

6. The following table gives some information about four drugs.

Drug	Type of action	Addictive?
anabolic steroids	performance enhancer	no
aspirin	pain killer	no
barbiturate	depressant	yes
nicotine	stimulant	yes

(a) Which drug shown in the table is found in cigarette smoke? [1]
(b) Two of the drugs in the table are addictive. What does this mean? [2]
(c) What effect would barbiturates have on the nervous system? [1]
(d) Barbiturates are a class B drug. Why are drugs put into different classes? [2]

7. Look back at the How Science Works article about drug testing.

(a) Suggest two reasons why drugs are tested on animals. [2]
(b) Explain why placebos are used when testing drugs. [2]
(c) In the 1960s, a drug called thalidomide was given to pregnant women. The drug had been tested on animals and seemed to be safe. Soon it became clear that the drug was causing the women to give birth to children with birth defects. Some animal rights campaigners say that this shows that drug testing on animals is a waste of time.

Write an argument for and against this view. [4]

The following topics are covered in this chapter:

- Genes and chromosomes
- Manipulating genes
- Variation and evolution

3.1 Genes and chromosomes

What is a gene?

OCR A   B1.1
OCR B   B1g
AQA   B1.11.6
EDEXCEL 360 B1a2

Most cells contain a nucleus that controls all of the chemical reactions that go on in the cell. Nuclei can do this because they contain the **genetic material**. Genetic material controls the characteristics of an organism and is passed on from one generation to the next. The genetic material is made up of structures called **chromosomes**. They are made up of a chemical called deoxyribonucleic acid or **DNA**. The DNA controls the cell by coding for the making of proteins, such as enzymes. The enzymes control all the chemical reactions taking place in the cell.

 KEY POINT   A gene is part of a chromosome that codes for one particular protein.

DNA codes for the proteins it makes by the order of four chemicals called bases. They are given the letters **A**, **C**, **G** and **T**.

Each one of our cells has about 30 000 genes.

Although the cells in one person have the same genes, they do not use them all. For example, a liver cell will use some genes and a cheek cell will use different genes.

The Human Genome Project has mapped all the genes on our chromosomes.

By controlling cells, genes therefore control all the characteristics of an organism. Different organisms have different numbers of genes and different numbers of chromosomes. In most organisms that reproduce by sexual reproduction, the chromosomes can be arranged in pairs. This is because one of each pair comes from each parent.

Chromosomes and reproduction

OCR A   B1.4
OCR B   B1g
AQA   B1.11.6
EDEXCEL 360 B1a2

No living organism can live for ever, so there is a need to reproduce. There are two main methods that organisms use to reproduce: sexual and asexual.

Sexual reproduction

Sexual reproduction involves the passing on of genes from two parents to the offspring. This is why we often look a little like both of our parents. The genes are passed on in the sex cell or gametes.

> **KEY POINT**
> Sexual reproduction involves the joining together of male and female sex cells or gametes.

In humans each body cell has 46 chromosomes in 23 pairs. This means that when the male sex cells (sperm) are made they need to have 23 chromosomes one from each pair. The female gametes (eggs) also need 23 chromosomes. When they join at fertilisation they produce a cell called a **zygote** that has 46 chromosomes again. This will grow into an **embryo** and become a baby.

> Identical twins are made when the embryo splits into two, so they have the same genes as each other.

This also means that the offspring that are produced from sexual reproduction are all different because they have different combinations of chromosomes from their mother and father.

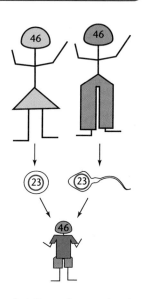

Fig. 3.1 Sexual reproduction.

Asexual reproduction

> Gardeners often use asexual reproduction to copy plants – they know what the offspring will look like.

Bacteria, plants and some animals can reproduce asexually. This needs only one parent and does not involve sex cells joining.
All the offspring that are made are genetically identical to the parent.
Different organisms have different ways of reproducing asexually but one example is the spider plant. This grows new plantlets on the end of long shoots.

Fig. 3.2 A spider plant uses asexual reproduction to reproduce.

Sex determination

Humans have 23 pairs of **chromosomes**. The chromosomes of one of these pairs are called the sex chromosomes because they carry the genes that determine the sex of the person.

> **KEY POINT**
> There are two kinds of sex chromosome. One is called X and one is called Y.

- Females have **two X chromosomes** and are **XX**.
- Males have an **X and a Y chromosome** and are **XY**.

> It is possible to separate X and Y sperm in the laboratory and so choose the sex of a baby.

This means that females produce ova that contain single X chromosomes. Males produce sperm, half of which contain a Y chromosome and half of which contain an X chromosome.

The reason why the sex chromosomes determine the sex of a person is due to a single gene on the Y chromosome. This gene causes the production of testes rather than ovaries and so the male sex hormone testosterone is made.

	X	Y
X	XX	XY
X	XX	XY

Passing on genes

Because we have two copies of each chromosome in our cells (one from each parent) this means that we have two copies of each gene.

> **KEY POINT**
>
> **A copy of a gene is called an allele.**

Sometimes the two alleles are the same but sometimes they are different. A good example of this is tongue rolling. This is controlled by a single gene and there are two alleles of the gene, one that says roll and the other that says do not roll.

> The people who are non-rollers must have two recessive alleles.

But even if we have two alleles, if one says roll and the other says do not roll, then a person can still roll their tongue. This is because the allele for rolling is **dominant** and the non-rolling allele is **recessive**.

> **KEY POINT**
>
> Some important words that you need to know:
> - **homozygous** means that both alleles are the same
> - **heterozygous** means that the two alleles are different.

How to work out the results of a cross

We usually give the alleles letters, with the dominant allele having a capital letter. For example, T = tongue rolling and t = non-rolling.

Let us assume that mum cannot roll her tongue but dad can.

Both of dad's alleles are T so he is homozygous.

The cross is usually drawn out like this:

> In this cross all the children can roll their tongue.

		mum	
		t	t
dad	T	Tt	Tt
	T	Tt	Tt

all are tongue rollers

If both mum and dad are heterozygous the children that they can produce will be different:

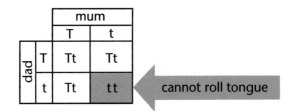

		mum	
		T	t
dad	T	Tt	Tt
	t	Tt	tt

cannot roll tongue

Inherited disorders

Many genetic disorders are caused by certain alleles. These can be passed on from mother or father to the baby and lead to the baby having the disorder. Examples of these disorders are **cystic fibrosis** and **Huntington's disorder**.

> **KEY POINT**
> Cystic fibrosis is caused by a recessive allele and Huntington's is caused by a dominant allele.

People with these disorders become ill:

Cystic fibrosis	Huntington's disorder
mucus collects in the lung	muscle twitching
breathing is difficult	loss of memory
food is not properly digested	difficulty in controlling movements

By looking at family trees of these genetic disorders and drawing genetic diagrams like the one for tongue rolling, it is possible for people to know the chance of them having a child with a genetic disorder. This may leave them with a difficult decision to make as to whether to have children or not.

3.2 Manipulating genes

Genetic engineering

OCR A B1.3
AQA B1.11.6
EDEXCEL 360 B1a2

All living organisms use the same language of DNA. The four 'letters' **A**, **G**, **C** and **T** are the same in all living things. Therefore a gene from one organism can be removed and placed in a totally different organism where it will continue to carry out its function.

> **KEY POINT**
> Moving a gene from one organism to another is called genetic engineering.

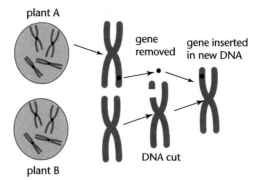

plant A

gene removed

gene inserted in new DNA

DNA cut

plant B

Fig. 3.3 Genetic engineering.

New genetically modified (GM) plants can be made in this way that may be more resistant to disease or produce a higher yield.

People often have different views about genetic engineering:
- Some people think that genetic engineering is against 'God and Nature' and is potentially dangerous.
- Some people think that genetic engineering will provide massive benefits to humans, such as better food and less disease.

There is also the possibility that genetic engineering may be used to treat genetic disorders like cystic fibrosis. Scientists are trying to replace the genes in people who have the disorder with working genes.

> **KEY POINT**
> **Using genetic engineering to treat genetic disorders is called gene therapy.**

Again this is quite controversial because it could be used to change the genes of embryos. People are worried that it might be used to produce 'designer babies'.

Cloning

OCR A — B1.4
AQA — B1.11.6
EDEXCEL 360 — B1a2

Asexual reproduction produces organisms that have the same genes as the parent.

This means that they will be very similar.

> **KEY POINT**
> **Genetically identical individuals are called clones.**

Many plants such as the spider plant do this naturally and it is easy for a gardener to **take cuttings** to make identical plants.

Modern methods involve **tissue culture**, which uses small groups of cells taken from plants to grow new plants.

Cloning animals is much harder to do. Two main methods are used:
- **Cloning embryos.** Embryos are split up at an early stage and the cells are put into host mothers to grow.
- **Cloning adult cells.** The first mammal to be cloned from adult cells was Dolly the sheep.

Since Dolly was born other animals have been cloned and there has been much interest about cloning humans.

There could be two possible reasons for cloning humans:

- **Reproductive cloning** to make embryos for infertile couples.
- **Therapeutic cloning** to produce embryos that can be used to treat diseases.

The use of embryos to treat disease is possible due to the discovery of stem cells.

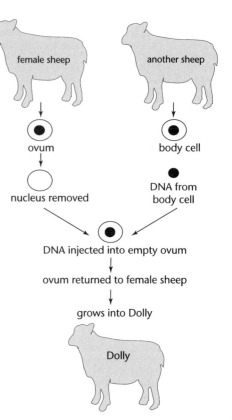

Fig. 3.4 How Dolly was cloned.

  Stem cells are cells that have the ability to divide to make any of the different tissues in the body.

It is much easier to find stem cells in an embryo and scientists think that they could be used to repair damage such as injuries to the spinal cord. There are therefore many different views about cloning:

Both infertility and genetic diseases cause much pain and distress. I think that we should be able to use cloning to treat these problems.

It is not right to clone people because clones are not true individuals and it is not right to destroy embryos to supply stem cells

3.3 Variation and evolution

Variation

OCR A   B3.1
OCR B   B1h

Children born from the same parents all look slightly different. We call these differences 'variation':

- **Inherited or genetic** – some variation is inherited from our parents in our genes.
- **Environmental** – some variation is a result of our environment.

Often our characteristics are a result of both our genes and our environment. Here are some examples:

Inherited	Environmental	Inherited and environmental
earlobe shape	scars	intelligence
eye colour	spoken language	body mass
nose shape		height

> Rare cases of identical twins that have been brought up separately can provide good evidence to investigate this.

The genes provide a height and weight range into which we will fit, and how much we eat determines where in that range we will be.

Scientists have argued for many years whether 'nature' or 'nurture' (inheritance or environment) is responsible for characteristics such as intelligence, sporting ability and health.

Because the baby can receive any one of the 23 pairs from mum and any one of the 23 pairs from dad, the number of possible gene combinations is enormous. This new mixture of genetic information produces a great deal of variation in the offspring. This just mixes genes up in different combinations but the only way that new genes can be made is by **mutation**.

> **KEY POINT**   **A mutation is a random change in a gene.**

A gene mutation occurs when one of the chemical 'letters' in DNA is changed. When this happens, it is most unlikely to benefit the organism. Think what would happen if you made random changes to a few of the letters on this page. It is most likely to produce gibberish and very unlikely to make any sense at all.

- If a mutation occurs in a gamete, the offspring may develop abnormally and could pass the mutation on to their own offspring.
- If a mutation occurs in a body cell, it could start to multiply out of control – this is **cancer**.

But very occasionally, a mutation may be useful, and without mutations we would not be here.

Here are some causes of mutations:

- **radiation**
- **UV** in sunlight
- **X-rays**
- **chemical mutagens**
  – as found in cigarettes.

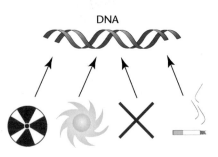

DNA

Fig. 3.5 Causes of mutations.

Evolution

OCR A B3.1
OCR B B2f
AQA B1.11.7

Most scientists now think that life on Earth started about 3500 million years ago. But how life started and why there is such a great variety of organisms have been questions that people have argued over for a long time.

In the 1800s, scientists started to find **fossils** of many different animals and plants.

Records of organisms can be preserved in different ways:

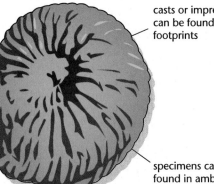

hard body parts get covered in sediment and gradually harden with minerals to become fossils

casts or impressions can be found, such as footprints

specimens can be found in amber, peat bogs, tar pits and ice

Fig. 3.6 Evidence from dead organisms.

Many people at that time believed in **creation**. They said that organisms were created as they exist now by God. However, scientists found fossils of organisms such as **dinosaurs** that are not alive today. Some people started to believe the idea that species of organisms could gradually change.

 KEY POINT Evolution is the gradual change in a species over a long period of time.

People argued against evolution by saying that the fossil record was incomplete and that the fossils of some organisms are missing.
Possible reasons for this are:

- Some body parts do not fossilise.
- Conditions have to be just right for fossils to be made.
- Many fossils may not yet have been found.

The problem for the believers in evolution was that at first they could not explain how the gradual changes happened.

Charles Darwin and natural selection

Many of these observations were made on the Galapagos Islands off the coast of South America.

Charles Darwin (1809–1882) was a naturalist on board a ship called the HMS *Beagle.* His job was to make a record of the wildlife seen at the places the ship visited.

On his travels, Darwin noticed four things:

● Organisms often produce **large numbers** of offspring.
● Population numbers usually remain constant over long time periods.
● Organisms are all slightly **different**: they show **variation**.
● This variation can be inherited from their parents.

Darwin used these four simple observations to come up with a theory for how evolution could have happened.

Darwin said that all organisms are slightly different and some are **better suited** to the environment than others. These organisms are more likely to survive and reproduce.

They will pass on these characteristics and over long periods of time the species will change.

> **KEY POINT**   **Darwin called this theory natural selection.**

Darwin was rather worried about publishing his ideas. When he finally published them they caused much controversy. Many people were very religious and believed in creation. It took many years before Darwin's theory was generally accepted.

Natural selection in action

Because natural selection takes a long time to produce changes, it is very difficult to see it happening. One of the first examples to be seen was the peppered moth.

Fig. 3.7 Light and dark peppered moths.

This moth is usually light coloured, but after the Industrial Revolution a black type became common in polluted areas. This can be explained by **natural selection**:

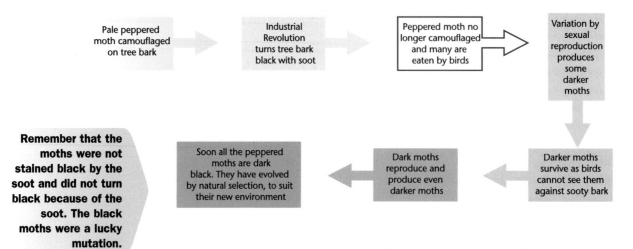

| Pale peppered moth camouflaged on tree bark | → | Industrial Revolution turns tree bark black with soot | → | Peppered moth no longer camouflaged and many are eaten by birds | ⇒ | Variation by sexual reproduction produces some darker moths |

| Soon all the peppered moths are dark black. They have evolved by natural selection, to suit their new environment | ← | Dark moths reproduce and produce even darker moths | ← | Darker moths survive as birds cannot see them against sooty bark |

Remember that the moths were not stained black by the soot and did not turn black because of the soot. The black moths were a lucky mutation.

MRSA bacteria are discussed on page 29.

Other examples that can be explained by natural selection include:

- Rats becoming resistant to the rat poison warfarin.
- Bacteria becoming resistant to antibiotics.

How did life start?

OCR A   B3.1

The atmosphere at the time contained different gases from those in the atmosphere now. The electric spark copies the action of lightning.

Once people accepted natural selection and evolution, scientists tried to work out how life might have started.

Experiments have shown that molecules that are important to life can be made in conditions that are similar to those found on the earth 3500 million years ago.

Molecules that make up DNA can be produced and DNA is found in all living cells. It can copy itself and carry genetic codes, so it is vital to life.

Some scientists think that other molecules may have come from space, perhaps from comets.

Once living cells were produced, natural selection could produce the variety of life that is alive today.

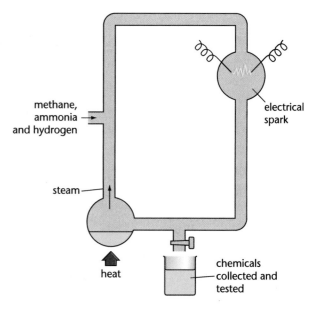

Fig. 3.8 How to create molecules important to life.

HOW SCIENCE WORKS

Evolution – explaining the facts

Living on the Earth are about five million different types of living organisms.

During humankind's time on Earth many people have put forward ideas to try and explain how this has come about.

Three different ideas are shown:

Creation theory says that the earth and life on it were created by God as described in the bible. Only small changes have happened since creation and no new species have been created.

Darwin said that all organisms were slightly different. Those organisms that were better suited would pass on their characteristics and so the population would gradually

A French scientist called **Lamarck** said that organisms were changed by their environment during their life. They then passed on the new characteristics and so the population would change.

Each of these ideas is a **theory** that explains **data** that is known at the time. As different data becomes known, then people often start to accept different theories. Since the discovery of many different extinct fossils and the dating of rock many people think that the creation theory cannot explain the data.

Most scientists now think that Lamarck's theory is wrong because we now know that characteristics are passed on in our genes and genes are not usually altered by the environment.

Most people now accept Darwin's theory because it best explains all the data that has been discovered. But it is only a theory, not fact.

HOW SCIENCE WORKS

Manipulating genes – right or wrong?

In 1990, two American girls took part in the world's first gene therapy experiment to try to repair their immune systems. The gene therapy involved putting genes into their white blood cells, using a virus to inject the genes. The normal virus genes had first been removed. The experiment was successful. Without those new genes, it was unlikely they could have survived. Their survival signalled the possible start of using gene therapy to cure many life-threatening genetic disorders.

But after 5000 patients had participated in 350 trials, things began to go wrong. First, in 1999, a young man being treated died of a massive immune reaction to the gene treatment. Three years later, a French baby developed leukaemia as a by-product of the treatment. Experts say the virus that inserted the genes mistakenly turned on a cancer-causing gene.

Scientists are now learning from these mistakes and further research into gene therapy is taking place. Cystic fibrosis is the most common inherited disease in the Western world. In 1989, the gene that causes cystic fibrosis was discovered. This caused excitement that this disorder could be treated with gene therapy.

Scientists have tried for ten years to put genes into the cells of the lungs to cure cystic fibrosis, but with little success. It seems to work in cells in a test tube but not in the body.

Now a new idea is about to be tried. This involves using stem cells. These are cells that can develop into different types of cells. They can be taken from the body of the patient and then the normal gene can be put inside them. The cells can then be put back into the patient's lungs where they divide to make normal lung cells.

The possible use of gene therapy and stem cells raises some ethical issues. To make the stem cells involves the destroying of embryos.

Some people think that this is not acceptable. Others think that it is wrong to stop people having a treatment that might cure their disease.

Exam practice questions

1. The theory put forward by Charles Darwin to explain how evolution could have occurred is called:
   - (a) selective breeding
   - (b) cloning
   - (c) natural selection
   - (d) genetic engineering [1]

2. Which row, A, B, C or D, in the following table contains the correct characteristics? [2]

	Characteristic is controlled by		
	Genes only	Environment only	Both
A	height	eye colour	body mass
B	body mass	scars	earlobe shape
C	scars	body mass	height
D	blood group	spoken language	intelligence

3. A genetically identical copy of an individual is called a:
   - (a) clone
   - (b) mutation
   - (c) gamete
   - (d) zygote

4. Finish the sentences by writing the correct word in the gap. Choose your answers from this list.

   chromosomes   cytoplasm   DNA   nucleus   proteins   sugar

   The genes in cells are found in the _____.
   These genes are on long strands called _____ .
   They are made of a chemical called _____.
   The genes control the cell by describing which _____ the cell should make. [4]

5. The diagram shows the chromosomes present in a human skin cell.

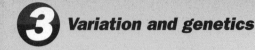

Exam practice questions

(a)   How would these chromosomes be different if they were from a sex cell and not a skin cell?   [2]

(b)   The cell is from a male. How can you tell this?   [2]

6.   Jackie has just had a baby. Leroy is the father. The doctor has told them that their baby has cystic fibrosis even though neither Jackie nor Leroy has the disorder.

(a)   Is the copy of the gene (allele) that causes cystic fibrosis dominant or recessive? How can you tell?   [2]

(b)   Fill in the boxes below to show how Jackie and Leroy could have had a baby that has cystic fibrosis. Jackie's gametes have been done for you.   [3]

	Jackie	
gametes	**F**	**f**

Leroy

(c)   If Jackie and Leroy had another baby, what would be the chance of it having cystic fibrosis?   [1]

(d)   The doctor tells Jackie that if she gets pregnant again they could have the foetus tested to see if it had cystic fibrosis. Suggest why it may be a difficult decision for Jackie to decide whether to be tested or not.   [2]

7.   Look back at the How Science Works article about evolution.

When Darwin returned from his journey on the HMS *Beagle* he wrote down his ideas about how evolution may have happened.

(a)   Suggest why Darwin was worried about publishing these ideas.   [2]

(b)   How did the discovery of fossils of extinct animals help to persuade some people that Darwin's ideas were right?   [2]

(c)   What is the difference between a theory such as natural selection and data such as fossils?   [3]

4 Organisms and the environment

The following topics are covered in this chapter:

- **The variety of life**
- **Living together**
- **Human impact on the environment**
- **Conservation**

4.1 The variety of life

Classifying organisms

OCR B   B1b
EDEXCEL 360   B1a1

Humans have been classifying organisms into groups ever since they started studying them.

- This makes it convenient when trying to identify an unknown organism.
- It also tells us something about how closely organisms are related and about their evolution.

The modern system that we use puts organisms into a system of smaller and smaller groups. **Kingdoms** are the largest groups. The kingdoms are divided into smaller and smaller groups until the smallest group is formed, called a **species**.

Members of a species are very similar, but how do we know if two similar animals are the same species?

> **KEY POINT**   Members of the same species can breed with each other to produce fertile **offspring**.

Animals such as mules are called hybrids and are often very strong.

Horses and donkeys are different species because, although they can mate and produce a mule, mules are **infertile**.

horse   donkey   mule

Fig. 4.1 Horse + donkey = mule!

Deciding on groups

OCR B   B1b
EDEXCEL 360   B1a1

The first step in classifying an organism is to put it into a **kingdom**.
Two of the kingdoms are the **plant and animal kingdoms**.

> Most of the differences between plants and animals are due to the way that they feed.

Plants	Animals
make their own food	take their food in ready-made
move by growing	move the whole of their body
grow in a spreading shape	have a compact shape

This makes it easy to classify most plants and animals but there are problems.
Fungi grow rather like plants but cannot make their own food and therefore
they are put in a separate kingdom.

Some microscopic organisms like *Euglena* can make their own food like plants
but in the dark start feeding like an animal. They are put in another kingdom
called **Protoctista**.

Bacteria are so different in structure from other organisms that they are put in
a separate kingdom. This makes five kingdoms in all.

Once an organism is put into the animal kingdom it can be put into the
vertebrate group or one of several invertebrate groups.

The vertebrate group is divided into five different classes.

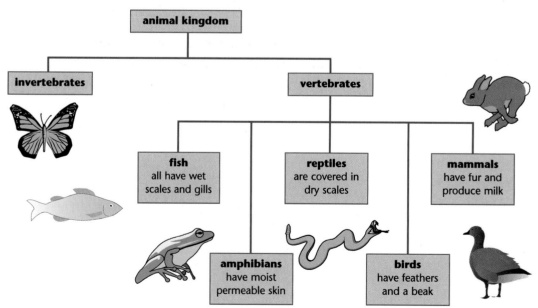

Fig. 4.2 The animal kingdom.

Naming organisms

Organisms are often known by different names in different countries or even in different parts of the same country. All organisms are therefore given a **scientific name** that is used by scientists in every country. This avoids confusion.

> **KEY POINT** The scientific system of naming organisms is called the **binomial system**.

Each name has two parts. The first part is the name of the genus (the group above species). The second part of the name is the species. For example:

> The binomial name is always typed in italics and the genus starts with a capital letter but the species does not.

Lion is *Panthera leo*   Tiger is *Panthera tigris*

These animals are in the same genus but are different species.

4.2 *Living together*

Where do organisms live?

Different organisms live in different environments.
- The place where an organism lives is called its **habitat.**
- All the organisms of one type living in a habitat are called a **population**.
- All the populations in a habitat are a **community**.
- An ecosystem is all the living and non-living things in a **habitat**.

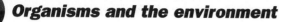

Our planet has a range of different ecosystems. Some of these are natural, such as woodland and lakes. Others are artificial and have been created by people, such as fish farms, greenhouses and fields of crops. Artificial ecosystems usually have less variety of organisms living there (less biodiversity). This may be caused by the use of chemicals such as weed killers, pesticides and fertilisers.

It is possible to investigate where organisms live by using various devices.

> **The more samples that you take in an area then the more accurate the estimate of the whole area will be.**

A **quadrat** is a small square that is put on the ground to take a sample of a large area. The number of organisms in the quadrat can be counted and the size of the population in the whole area can then be estimated.

Quadrats are often used to study plants but devices such as pooters, nets and pit-fall traps can be used to sample animal populations.

Competition and adaptation

OCR A   B3.4
OCR B   B2d
AQA   B1.11.5
EDEXCEL 360   B1a1

There are many different types of organisms living together in a habitat and many of them are after the same things.

> **KEY POINT**   This struggle for resources is called competition.

The more similar the organisms, the greater the competition.
Plants usually compete for:
- **light** for photosynthesis
- **water**
- **minerals**.

> **Organisms of the same species are more likely to compete with each other because they have similar needs.**

Animals usually compete for:
- **food** to eat
- **water** to drink
- **mates** to reproduce with
- **shelter**.

Because there is constant competition between organisms, the best suited to living in the habitat survive. Over many generations the organisms have became suited to their environment.

> **KEY POINT**   The features that make organisms well suited to their environment are called adaptations.

Habitats such as the Arctic and deserts are difficult places to live because of the extreme conditions found there. Animals and plants have special adaptations so they can survive:

Polar bears have:	Cacti have:	Camels have:
a large body that holds heat	leaves that are just spines to reduce surface area	a hump that stores food as fat
thick insulating fur	deep or widespread roots	thick fur on top of the body for shade
a thick layer of fat under the skin	water stored in the stem	thin fur on the rest of the body
white fur that is a poor radiator of heat and provides camouflage		

Flowering plants also show adaptations. Some are adapted to being **pollinated** by insects and some are pollinated by wind.

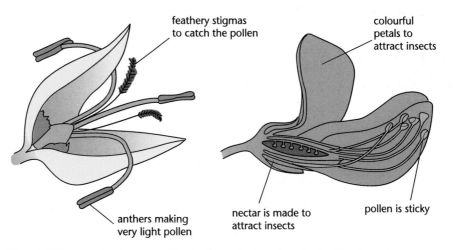

feathery stigmas to catch the pollen

colourful petals to attract insects

anthers making very light pollen

nectar is made to attract insects

pollen is sticky

Fig. 4.3 Flower structure – pollinated by wind and pollinated by insects.

Relationships

Predator and prey

Organisms form different types of relationships with other organisms in their habitat.

One of the most common is that of predator and prey.

> **KEY POINT** A predator **hunts and kills another animal for food. The animal that is eaten is called the prey.**

The numbers of predators and prey in a habitat will vary and will affect each other. The size of the two populations can be plotted on a graph called a **predator–prey graph**.

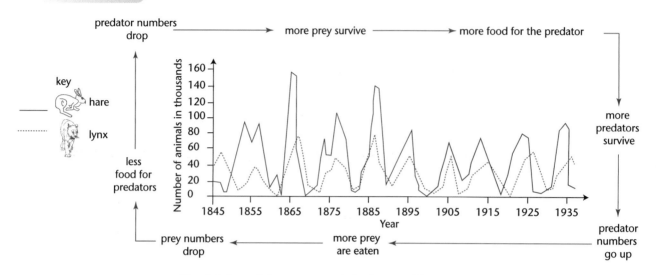

predator numbers drop → more prey survive → more food for the predator

key

 hare

lynx

less food for predators

more predators survive

prey numbers drop ← more prey are eaten ← predator numbers go up

Fig. 4.4 A predator–prey graph for lynx and hares.

Parasite and host

Sometimes one organism may not kill another organism but may take food from it while it is alive.

> **KEY POINT**
> A parasite lives on or in another living organism called the host, causing it harm.

Many diseases such as **malaria** are caused by parasites feeding on a host. The parasite in malaria is a single-celled organism that feeds on humans, who are the host. The organism is injected into the blood stream by a mosquito. This is also acting as a parasite but it known as a **vector** for malaria because it spreads the disease-causing organism without being affected by it.

Mutualism

Instead of trying to eat each other, some different types of organisms work together.

> **KEY POINT**
> When two organisms of different species work together so that both gain, this is called mutualism.

Examples of this type of relationship are:
- Oxpeckers and buffalos: the oxpecker birds eat the parasites on the backs of the buffalo.
- Nitrogen-fixing bacteria in the roots of pea plants: the bacteria give the plants nitrates and they gain sugars from the plant.

Food production

OCR B   B2c

Plants make their own food by a process called **photosynthesis**. They take in **carbon dioxide** and **water** and turn it into **sugars**, releasing **oxygen** as a waste product. The process needs the **energy** from sunlight and this is trapped by the green pigment **chlorophyll**.

> **Look back at the equation for respiration on page 11. It is the reverse of this equation.**

carbon dioxide + water $\xrightarrow[\text{chlorophyll}]{\text{light}}$ glucose + oxygen

$$6CO_2 + 6H_2O \rightarrow C_6H_{12}O_6 + 6O_2$$

Once plants have made sugars such as glucose they can convert the glucose into many different things:

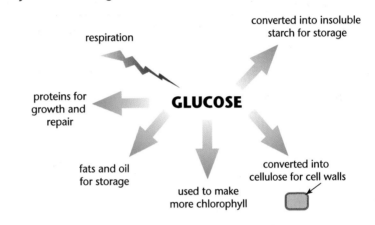

Limiting factors

The rate of photosynthesis can be increased by providing:

- more **light**
- more **carbon dioxide**
- an **optimum temperature**.

Any of these factors can be limiting factors.

> **KEY POINT** A limiting factor **is something that controls how fast a reaction will occur.**

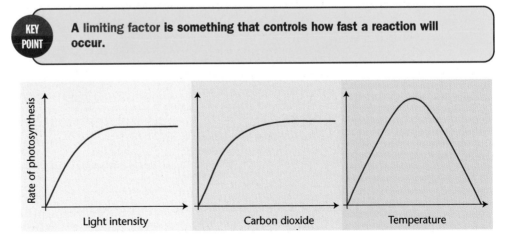

Fig. 4.5 Limiting factors for photosynthesis.

If more light is provided, it will increase photosynthesis because more energy is available. After a certain point something else will limit the rate.

More carbon dioxide will again increase the rate up to a point because more raw materials are present. Increasing temperature will make enzymes work faster but high temperatures prevent enzymes from working.

Respiration and photosynthesis

Many people think that plants respire at night and photosynthesise during the day.

In fact, plants carry out respiration all the time. Fortunately for us, during the day they photosynthesise much faster than they respire. This means that they make enough oxygen and food for us.

Energy transfer

EDEXCEL 360 B1a1

A **food chain** shows how food passes through a community of organisms. It enters the food chain as **sunlight** and is trapped by the **producers**. These are the **green plants**. The energy then passes from organism to organism as they eat each other.

Often the waste from one food chain can be used by decomposers to start another chain.

The mass of all the organisms at each step of the food chain can be measured. This can be used to draw a diagram that is similar to a **pyramid of numbers**. The difference is that the area of each box represents the mass of all the organisms, not the number.

> **KEY POINT**
>
> This type of diagram is called a pyramid of biomass.

The reason that a pyramid of biomass is shaped like a pyramid is that energy is lost from the food chain in different ways as the food is passed along.

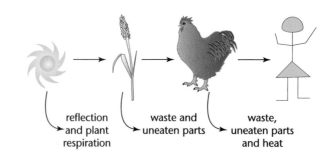

Fig. 4.6 Where energy is wasted in the food chain.

This means that less energy is lost if a person eats plant products than if they eat meat. This is because the food only goes through one transfer rather than two.

4.3 Human impact on the environment

Growth in populations

The human population on Earth has been increasing for a long time but it is now going up more rapidly than ever. The rate of increase is increasing and this is called **exponential growth**.

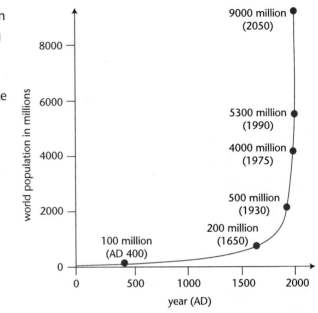

Fig. 4.7 Population growth.

This increase in the population is having a number of effects on the environment:

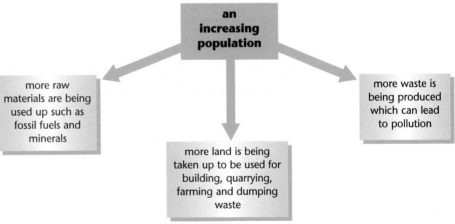

Pollution

Modern methods of food production and the increasing demand for energy have caused many different types of **pollution**.

 KEY POINT Pollution is the release of substances into the environment that harm organisms.

The table shows some of the main polluting substances that are being released into the environment.

Polluting substance	Main source	Effects on the environment
carbon dioxide	burning fossil fuels	greenhouse effect
CFCs	fridges and aerosols	destroy the ozone layer
fertilisers	intensive farming	pollute rivers and lakes
herbicides	intensive farming	some cause mutations
methane	cattle and rice fields	greenhouse effect
sewage	human and farm waste	pollutes rivers and lakes
sulphur dioxide	burning fossil fuels	acid rain

The **greenhouse effect** is caused by a build-up of certain gases, such as carbon dioxide and methane, in the atmosphere. These gases trap the heat rays as they are radiated from the earth. This causes the Earth to warm up. This is similar to what happens in a greenhouse. This could lead to changes in the Earth's climate and a large rise in sea level.

Some scientists disagree over how severe the greenhouse effect is and what is actually causing it. See page 68.

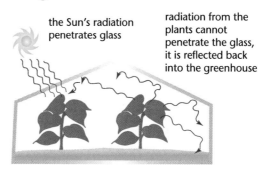

the Sun's radiation penetrates glass

radiation from the plants cannot penetrate the glass, it is reflected back into the greenhouse

Fig. 4.8 The greenhouse effect.

Acid rain is caused by the burning of fossil fuels that contain some **sulphur** (can also be spelt 'sulfur') impurities. This gives off sulphur dioxide, which dissolves in rainwater to form **sulphuric acid**. This falls as acid rain.

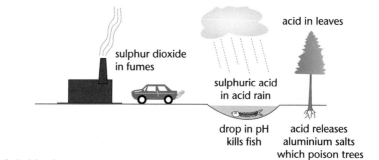

sulphur dioxide in fumes

acid in leaves

sulphuric acid in acid rain

drop in pH kills fish

acid releases aluminium salts which poison trees

Fig. 4.9 Acid rain.

Ozone depletion is caused by the release of chemicals such as **CFCs** which come from the breakdown of refrigerators and aerosol sprays. **Ozone** helps protect us from harmful UV radiation and so depletion may lead to more skin cancer.

Pollution indicators

Some organisms are more sensitive to pollution than others. If we look for these organisms, it can tell us how polluted an area is.

On land, lichens grow on trees and stone.

Some lichens are killed by lower levels of pollution than other types.

In water some animals, such as the rat-tailed maggot, can live in polluted water but others, such as mayfly larvae, live only in clean water.

Fig. 4.10 Lichens can tell us how polluted the environment is.

Fig. 4.11 Rat-tailed maggot.

Over-exploitation

Deforestation

As well as causing pollution, the increasing demands for food, land and timber have caused people to cut down large areas of forests.

> **KEY POINT**
>
> Cutting down large areas of trees is called deforestation.

Deforestation has led to:
- Less carbon dioxide being removed from the air by the trees and carbon dioxide being released when the wood is burnt.
- The destruction of habitats that contain some rare species.

Overhunting and extinction

Some animals have been hunted until their numbers have been dramatically reduced.

Many species of whales have been hunted for food, oil and other substances. Their numbers now are very low and people are trying to protect them.

Other organisms have not been so lucky. Their numbers have decreased so far that they have completely died out. This is called **extinction**. Organisms do become extinct naturally, but people have often increased the rate either directly or indirectly by:
- changing the climate
- destroying habitats
- pollution
- competition
- over-hunting.

Other organisms are at risk of becoming extinct and are **endangered**.

Fig. 4.12 Extinct and endangered animals.

4.4 Conservation

Biodiversity

OCR A B3.4
OCR B B2h
AQA   B1.11.8

Many people believe that is wrong for humans to damage natural habitats and cause the death of animals and plants. They believe that it is important to keep a wide variety of different animals and plants alive.

> **KEY POINT**
>
> The variety of different organisms that are living is called biodiversity.

There are many reasons given for trying to maintain biodiversity:

 It is natural for organisms to become extinct but in many cases humans are speeding up this process.

- Losing organisms may have unexpected effects on the environment, such as the erosion caused by deforestation.
- Losing organisms may have effects on other organisms in their food web.
- Some organisms may prove to be useful in the future, for breeding, producing drugs or for their genes.

Looking after the environment

OCR B B2h

To be able to save habitats and organisms, people have set up many different schemes.

> **KEY POINT**
>
> The attempt to preserve habitats and keep species alive is called conservation.

Many zoos are now more involved in conservation rather than just showing animals to people.

There are a number of different ways that conservation programmes can work:

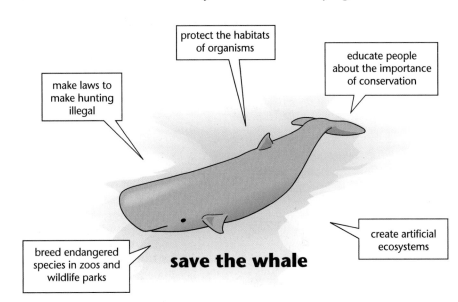

protect the habitats of organisms

educate people about the importance of conservation

make laws to make hunting illegal

create artificial ecosystems

breed endangered species in zoos and wildlife parks

save the whale

Fig. 4.13 How to save an endangered species.

Sustainable development

OCR A   B3.4
OCR B   B2h
AQA   B1.11.8

If the human population is going to continue to increase, it is important that we meet the demand for food and energy without causing pollution or over-exploitation.

 KEY POINT   **Providing for the increasing population without using up resources or causing pollution is called sustainable development.**

It has been quite difficult to get some countries to set targets on reducing carbon dioxide output.

To make sure that development is sustainable a lot of planning is needed at local, national and international levels.

In 1992, over 150 nations attended a meeting in Brazil called the Earth Summit. They agreed on ways in which countries could work together to achieve sustainable development.

They agreed to:
- reduce pollution from chemicals such as carbon dioxide. This can be done by cutting down on the waste of energy or by using sources of energy that do not produce carbon dioxide
- reduce hunting of certain animals, such as whales.

The document that they signed was called Agenda 21.
In 2002, a World Summit on Sustainable Development was held in Johannesburg, South Africa, to monitor progress.

HOW SCIENCE WORKS

Global warming – fact or theory?

The idea that the Earth's climate has changed over long periods in the past is a well-accepted **fact** among most people, but what may have caused this?

The theory that changes in our climate could be caused by changes in the carbon dioxide levels in the atmosphere was first put forward by a Swedish scientist called Svante Arrhenius in 1896.

We now call this theory the **greenhouse effect** or **global warming**.

But what is the evidence that global warming can be caused by increasing carbon dioxide levels?

Scientists have used several sets of **data** to try and provide evidence for this theory. One important set comes from the Antarctic Vostok ice core.

Vostok is a remote place in Antarctica where a hole has been drilled down into the ice over 3.3 kilometres deep. The ice core that has been removed contains bubbles of air that were trapped at different times many thousands of years ago. The gas has been analysed to tell us the levels of carbon dioxide and the temperature of the air in the past. Graphs have been plotted that look like this:

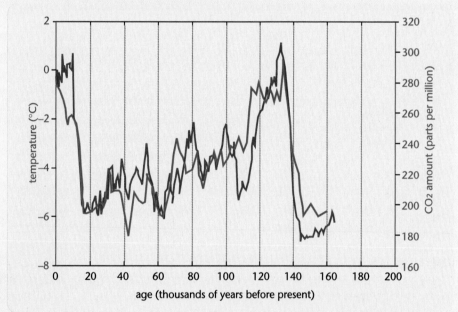

Looking at the graph, the levels of carbon dioxide and the temperature seem to follow a very similar pattern. Scientists say that there is a **correlation** between carbon dioxide levels and the temperature. This provides evidence for the theory.

However, this does not mean that one factor was **caused** by the other.

Even if one factor was caused by the other, it is possible to put forward two different theories:

- As the carbon dioxide levels increased, the suns rays were trapped and this caused the Earth to warm up.
- When the Earth warmed up, more animals survived and gave off more carbon dioxide.

Most scientists believe the first theory is correct, but theories are only ways of explaining data. They are not proven facts.

HOW SCIENCE WORKS

Making decisions about conservation is not always easy

Protecting whales has often been one of the main aims of conservation groups.

Records of whale hunting go back to 6000 BC. Whales have been hunted for their meat and for oil. The main countries involved in hunting have been Japan, Iceland and the United States.

In the 1800s, more modern methods of whaling reduced whale numbers dramatically and a number of whale species became endangered. Now an international group called the International Whaling Commission (IWC) exists to control whale hunting. Since 1985, commercial whaling has been banned, but some people object to this.

Norway has been hunting minke whales since 1993, killing about 600 every year.

Japan still hunts whales for 'scientific research'. According to official IWC figures, in the 2004–2005 whaling season, 601 minke whales were caught in coastal regions of Japan. Three sperm whales and 51 Bryde's whales were also taken.

In 2005, the research programme increased the quota of minke whales to 900 and, more controversially, added fin whales to the programme. This move has sparked a great deal of controversy among anti-whaling nations, in particular because fin whales are listed as endangered under the Convention on International Trade on Endangered Species. Starting in 2007, Japan plans to harvest up to 50 humpback whales and 50 fin whales annually.

Making decisions about protecting endangered species is not always very easy and clear cut. An example of this is the attempt to protect the grey whale.

A tribe of North American natives, the Makah, have hunted whales for thousands of years. They rely on the whales for food and the hunting is part of their culture. When whale numbers dropped, they had to give up hunting, and their culture and identity suffered. They now want to start hunting again on a small scale but the IWC has banned all hunting. Should they be allowed to hunt?

The arguments are still continuing.

- Should the Makah be allowed to kill a small number of whales a year to keep their culture alive?
- May this lead to other groups wanting to start hunting whales again?

Exam practice questions

1. Which of the following is not an effect of acid rain?
   (a) lowers the pH of lakes
   (b) poisons trees with aluminium salts
   (c) increases the temperature of the atmosphere
   (d) kills fish   [1]

2. Certain fungi grow on the roots of pine trees. The fungi take food from the tree roots. The fungi absorb minerals from the soil and some pass into the roots. The fungi are acting as:
   (a) parasites
   (b) mutualistic partners
   (c) predators
   (d) prey   [1]

3. A large number of greenfly live on an oak tree. This collection of greenfly is described as:
   (a) a community
   (b) an ecosystem
   (c) a class
   (d) a population   [1]

4. Complete the following sentences. Use words from this list.

   invertebrates   kingdoms   mammal   reptile   species   vertebrates

   Organisms are classified into five different _____.

   Animals with backbones are called _____. This group is then subdivided into five smaller groups; for example a mouse is a _____.

   The smallest group, which contains only one type of organism, is called a _____. [4]

5. Organisms are adapted to the environment that they live in.
   Explain how the following characteristics help the organism survive:
   (a) Camels store large amounts of fat in their humps.   [2]
   (b) Some cacti have deep roots that pass straight down whereas other types of
       cacti have shallow roots that spread out a long distance.   [3]
   (c) Polar bears are large animals with very small ears for the size of their body. [2]
   (d) The larvae of many insects do not feed on the same type of food as the
       adult insect.   [1]

Exam practice questions

6. A scientist wanted to estimate the number of greenfly and ladybirds in a large field.

    (a) Explain how he would use a quadrat to estimate the numbers of the two
        animals in the field.   **[3]**

    (b) The scientist estimated the numbers at different times of the year and
        plotted the results on a graph.

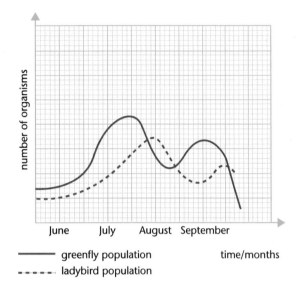

    greenfly population
    - - - - ladybird population
    time/months

    (i) Explain the shape of the scientist's graph   **[3]**

    (ii) What name is usually given to this type of graph?   **[1]**

7. Look back at the How Science Works article about whales.

    (a) Why is it difficult to be sure about the size of whale populations?   **[2]**

    (b) Discuss arguments for and against the Japanese approach to whale hunting. **[2]**

    (c) Why do some people think that the Makah tribe should be allowed to
        hunt a small number of whales?   **[2]**

The following topics are covered in this section:

- **Atoms**
- **Bonding**
- **Mixtures and compounds**
- **Equations**
- **The periodic table**

5.1 Atoms

What is inside an atom?

OCR A	C1.2
OCR B	C1
AQA	C1.12.1
EDEXCEL 360	C1a5

Matter is made up of tiny particles called **atoms**. Atoms are the basic building blocks of matter.

> **KEY POINT**
> Atoms are the smallest particles of an element that can still have the chemical properties of that element.

Atoms of all the elements are made up of three different particles: **protons**, **neutrons** and **electrons**.

These particles are smaller than the atoms themselves. We call them sub-atomic particles.

Particle	Symbol	Relative mass	Relative charge
proton	p	1	+1
electron	e	$\frac{1}{1840}$	−1
neutron	n	1	0 (neutral)

Most of an atom is empty space. If a football stadium represented an atom, the nucleus would be the size of the centre spot.

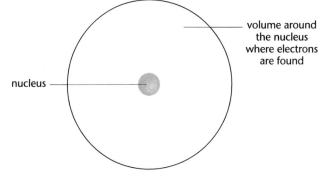

Fig. 5.1 The structure of an atom.

You can see in the diagram that an atom consists of a nucleus and electrons that move around the nucleus. The nucleus contains protons and neutrons. An atom has no overall charge because the number of protons is equal to the number of electrons.

Protons and neutrons are collectively called **nucleons**.

Elements

 KEY POINT An element is a pure substance that cannot be split up into two or more simpler substances by physical or chemical processes.

- An element is made of only one type of atom. Each element has its own type of atom. This means that the atoms of one element are not the same as the atoms of another element.
- There are 92 naturally occurring elements. Oxygen, hydrogen, nitrogen, sulphur, iron, copper, titanium and silver are examples of elements.
- Sugar is not an element. When heated, sugar decomposes to give water and carbon.
- Water is not an element. It can be broken down to give hydrogen and oxygen.
- Carbon, hydrogen and oxygen are elements because they cannot be further broken down into simpler substances.
- Chemists use chemical symbols as abbreviations for the names of elements. Each symbol consists of one or two letters.

See Figure 5.5 for the symbols of the elements.

KEY POINT All elements have their own symbols that may consist of one or two letters.

The atoms of an element are not always identical. For example, there are three types of hydrogen atoms. They all contain the same number of protons and electrons, but different numbers of neutrons. These are called **isotopes** of hydrogen.

KEY POINT Isotopes are atoms of the same element with the same number of protons but different numbers of neutrons.

The formation of ions

An atom becomes an ion if it loses or gains electrons.

KEY POINT An ion is a charged particle formed from an atom or a group of atoms by losing electrons or gaining electrons.

- Metals form positively charged ions (**cations**) whereas non-metals form negatively charged ions (**anions**).
- Positively charged ions are formed by the loss of electrons.
- Negatively charged ions are formed by the gain of electrons.

5.2 Bonding

- The joining of atoms together is called **bonding**.
- The arrangement of particles together is called a **structure**.

There are two main types of bonding:

- **ionic**
- **covalent**.

Ionic (or electrovalent) bond: the transfer of electrons

Sodium chloride

When sodium is heated and then placed in a gas jar of chlorine, the elements combine to form sodium chloride.

The sodium atom loses an electron to form a positively charged sodium ion. The chlorine atom gains an electron to form a negatively charged chloride ion. Positive sodium ions and negative chloride ions are attracted to one another by **electrostatic attraction** to form sodium chloride.

> Ionic bonding is sometimes called electrovalent bonding.

The strong electrostatic attraction that holds the oppositely charged ions together is called **ionic bonding**. Compounds that contain ionic bonds are called **ionic compounds**.

> **KEY POINT**
> The force of attraction that holds the sodium ions and chloride ions together is called ionic bonding.

Magnesium oxide

Magnesium burns in oxygen to form magnesium oxide. The magnesium atom loses two electrons to form a positively charged magnesium ion. The oxygen atom gains two electrons to form a negatively charged oxide ion. Positive magnesium ions and negative oxide ions are attracted to one another by electrostatic attraction to form magnesium oxide.

> **KEY POINT**
> Metals react with non-metals to form ionic compounds.

Some ions are made up of groups of atoms. These ions are called **polyatomic ions**. Examples of polyatomic ions are the ammonium ion (NH_4^+), carbonate ion (CO_3^{2-}) and sulfate(VI) ion (SO_4^{2-}).

Covalent bonding

OCR B   C1d
AQA   C1.12.1

> **KEY POINT** A covalent bond is formed by the sharing of a pair of electrons between two atoms.

When atoms combine by sharing electrons, molecules are formed.

> **KEY POINT** A molecule is a group of atoms held together by covalent bonds.

Chlorine molecule (Cl_2)

A chlorine atom shares a pair of electrons through covalent bonding.

- The sharing of two electrons is called a **single bond**.
- It is represented by a single line Cl–Cl.

Oxygen

The electronic configuration of an oxygen atom is (2,6). Each oxygen atom shares two pairs of electrons to form the covalent bond.

- The sharing of two pairs of electrons is called a **double bond**.
- It is represented by a double line O=O.

Molecules of compounds

A large number of compounds exist as covalent molecules. Covalent or molecular compounds are molecules made from two or more different atoms linked together by covalent bonding.

Water (H_2O), methane (CH_4), carbon dioxide (CO_2) and ammonia (NH_3) are examples of covalent compounds.

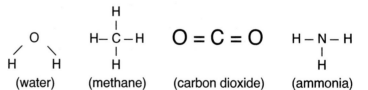

(water)   (methane)   (carbon dioxide)   (ammonia)

Models of these molecules are shown below.

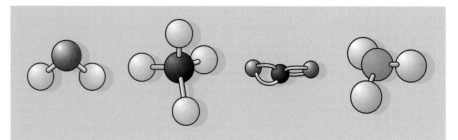

There are strong covalent bonds between atoms in a covalent molecule. Between the molecules are weak attractions called van der Waals' forces. For example, the covalent bonds between oxygen atoms in an oxygen molecule are strong. The bonds between oxygen molecules are weak. This helps to explain why oxygen is a gas.

Bonding, structure and properties

Bonding	Structure	Properties
Ionic	Giant structure, e.g. sodium chloride, magnesium oxide	High melting point Usually soluble in water but insoluble in organic solvents Conducts electricity when molten or when dissolved in water
Covalent	Molecular, e.g. chlorine, oxygen, methane	Usually gases or low boiling point liquids Usually insoluble in water but soluble in organic solvents Does not conduct electricity

If a substance conducts electricity when solid and when molten it must be a metal.

5.3 Mixtures and compounds

Mixtures

OCR A — C2.2
OCR B — C1d
AQA — C1.12.2
EDEXCEL 360 — C1b7

KEY POINT

Mixtures are formed when two substances are added together without chemical bonds being formed.

- Air is a mixture of oxygen, nitrogen, argon, carbon dioxide and other gases.
- Orange dye is a mixture of red dye and yellow dye.

Emulsions are mixtures. They contain tiny drops of one liquid spread evenly through a second liquid. An emulsifier is a substance that allows an emulsion to stay in a stable state. Emulsifiers are molecules that have two distinct ends. One end likes to be in water (hydrophilic) and the other likes to be in oil (hydrophobic or lipophilic).

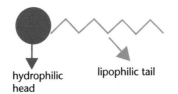
hydrophilic head   lipophilic tail

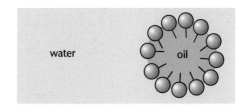
water   oil

Fig. 5.2 How an emulsifier works.

Emulsifiers cover the surface of the oil droplets in an oil-in-water emulsion. The hydrophobic end of the emulsifier is in water; the hydrophobic end is in oil. Milk is a natural emulsion. It is a mixture of fat droplets in water.

Emulsifiers are used in bread making. They help to give high-quality bread with an even texture. They stabilise the dough so that it can be processed in the bakery. They also reduce the rate at which bread goes stale.

no emulsifier   with emulsifier

Fig. 5.3 The texture of bread is much more even when an emulsifier is added.

Compounds

OCR A   C1.2
OCR B   C1a
AQA   C1.12.1
EDEXCEL 360   C1a5

KEY POINT

Compounds are formed when elements react together in fixed proportions.

- water (H_2O) is a compound of two atoms of hydrogen and one atom of oxygen
- magnesium oxide (MgO) is a compound of one atom of magnesium and one atom of oxygen.

The table gives the main differences between **mixtures** and **compounds**.

Mixtures, e.g. air, dyes, crude oil, emulsions	**Compounds**, e.g. water, sodium chloride, magnesium oxide
The different components of a mixture can be separated from one another by physical methods, e.g. filtration, distillation, chromatography	The elements in a compound can be separated only by chemical reactions or electrolysis
The physical and chemical properties of a mixture are the same as those of its components	The properties are different from those of the elements in the compound
No chemical reaction takes place when a mixture is formed and there is no energy change	A chemical reaction takes place when a compound is formed – usually there is an energy change, e.g. the reaction gets hot
The components of a mixture can be in any proportions	The elements in a compound are always combined in a fixed proportion by mass

Note that there are exceptions such as ammonia (NH$_3$), water (H$_2$O) and methane (CH$_4$).

Naming compounds

Compounds that are made up of two elements only have an '-ide' ending.

Compound	Formula	Elements in compound
carbon dioxide	CO$_2$	carbon and oxygen
zinc chloride	ZnCl$_2$	zinc and chlorine
hydrogen bromide	HBr	hydrogen and bromine
aluminium nitride	AlN	aluminium and nitrogen

Metal compounds that contain two or more elements, one of which is oxygen, have an '-ate' ending.

Note that there are exceptions, e.g. sodium hydroxide (NaOH).

Compound	Formula	Elements in compound
magnesium sulphate	MgSO$_4$	magnesium, sulphur and oxygen
calcium carbonate	CaCO$_3$	calcium, carbon and oxygen
potassium phosphate	K$_3$PO$_4$	potassium, phosphorus and oxygen
zinc nitrate	Zn(NO$_3$)$_2$	zinc, nitrogen and oxygen

Deducing the number of atoms in a formula

What can be deduced from the formula of calcium carbonate, which has a formula of CaCO$_3$?

- it contains the elements calcium, carbon and oxygen
- the various elements are found in the ratio Ca : H : O = 1 : 1 : 3
- how the atoms are linked together. The three oxygen atoms are linked to the carbon atom and not to the calcium atom.

The table shows some common substances, their formula and the number of atoms of each element.

Compound	Formula	Number of atoms of each element present
ammonia	NH$_3$	1 nitrogen, 3 hydrogen
carbon dioxide	CO$_2$	1 carbon, 2 oxygen
copper(II) carbonate	CuCO$_3$	1 copper, 1 carbon, 3 oxygen
ethanol	C$_2$H$_5$OH	2 carbon, 6 hydrogen, 1 oxygen
lead nitrate	Pb(NO$_3$)$_2$	1 lead, 2 nitrogen, 6 oxygen
methane	CH$_4$	1 carbon, 4 hydrogen
silicon dioxide	SiO$_2$	1 silicon, 2 oxygen
sodium chloride	NaCl	1 sodium, 1 chlorine
sugar	C$_6$H$_{12}$O$_6$	6 carbon, 12 hydrogen, 6 oxygen
sulphuric acid	H$_2$SO$_4$	2 hydrogen, 1 sulphur, 4 oxygen

Everyday examples of mixtures and compounds

Some foods may be eaten **raw,** for example fruits, nuts and vegetables such as carrots, peas and mangoes.

Other foods such as bread, eggs and potatoes are **cooked**.

There are different ways of cooking food. It can be:
- baked in an oven
- fried
- grilled
- microwaved
- steamed

During cooking
- a new substance is made
- the process is irreversible
- there is an energy change.

You may have made scones from self-raising flour, milk, salt and bicarbonate of soda. Once the scones have been made, it is impossible to get the ingredients back – a chemical change has taken place.

We cook food before we eat it for different reasons:
- the high temperature kills harmful microbes such as Salmonella
- to improve the flavour
- to improve the taste
- to improve the texture
- to make it easier to digest.

Two of the important ingredients in food are protein and carbohydrates (starch). Proteins are found in eggs and meat.

Albumen (spelt with an 'e') is another name for egg white. Albumin (with an 'i') is a protein.

The protein found in raw egg white is albumin. It exists with weak bonds between the protein molecules. Cooking an egg causes the albumin to permanently change its shape (**denature**). It allows it to form stronger bonds, which have the visible effect of stiffening the egg white and changing its colour from clear to white.

The proteins in meat undergo similar changes during cooking, also changing the appearance and texture of the meat.

The cell walls in vegetables such as potatoes are made of insoluble cellulose molecules. When potatoes are heated the cell walls break down, making the potato easier to digest.

Enzymes help to speed up this process.

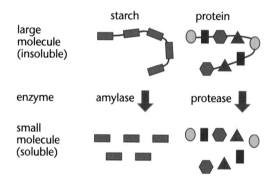

Fig. 5.4 Breaking down of large molecules.

Chromatography

> **KEY POINT**
>
> **Chromatography is a method of separating mixtures based on the different adsorptions of substances for paper and the solubilities of these substances in the solvents used.**

Many dyes are a mixture of other dyes. The dyes can be identified by using chromatography.

A pencil line is drawn on a piece of chromatography paper about 1 cm from the bottom of the paper. A small spot of the dye is put on the pencil line and allowed to dry. The chromatography paper is hung inside a glass tank containing a small volume of ethanol. (It is important that the dye spot is on the line, i.e. 1 cm above the level of **ethanol**. Ethanol acts as a solvent in this experiment.) The solvent travels up the paper, separating the dyes as it does so. When the solvent is nearly at the top of the chromatography paper, the paper is removed and allowed to dry. The various dyes making up the original dye can be seen (see Fig. 5.6).

You may have read about Sudan 1 being found in certain foods. All of the products that contained this chemical had to be removed from shops.

This is one of the methods used to identify colouring matter in food.

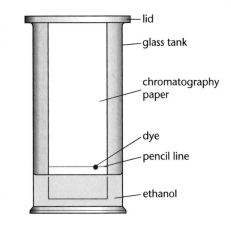

Fig. 5.5 Paper chromatography.

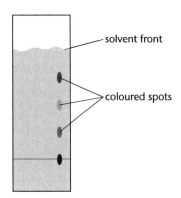

Fig. 5.6 The dye is separated into its three constituents.

5.4 Equations

Types of equation

A chemical equation is a 'shorthand' way of representing what occurs in a chemical reaction. It can be written in words or by using formulae.

Word equations

Hydrochloric acid reacts with magnesium to form magnesium chloride and hydrogen. The **word equation** for the reaction is

hydrochloric acid + magnesium $\rightarrow$ magnesium chloride + hydrogen

- The substances that react together are called the **reactants**.
- The new substances formed are called the **products**.
- In a chemical equation, the reactants are shown on the left-hand side of the equation and the products are shown on the right-hand side.
- The arrow ($\rightarrow$) means 'react to form'. It indicates that the reaction proceeds from left to right.

Symbol equations

Reactions can be summarised using **chemical symbols**. Follow these steps:

1. Write the word equation for the reaction, e.g.

hydrochloric acid + magnesium $\rightarrow$ magnesium chloride + hydrogen

2. Represent each of the reactants and products by their **correct** formulae

$HCl + Mg \rightarrow MgCl_2 + H_2$

3. Check the number of atoms of each element on both sides of the equation. If they are equal, the equation is balanced. If they are not equal, go on to step 4.

   In this case there are two chlorine atoms, two hydrogen atoms and one magnesium atom on the right-hand side and one chlorine atom, one hydrogen atom and one magnesium atom on the left-hand side. The numbers of magnesium atoms balance, but the numbers of hydrogen atoms and chlorine atoms do not balance.

4. Balance the equation by putting a '2' in **front** of the HCl. The equation is now balanced.

$$2HCl + Mg \rightarrow MgCl_2 + H_2$$

   There are two chlorine atoms and two hydrogen atoms on the right-hand side and two chlorine atom and two hydrogen atom on the left-hand side.

5. Now add **state symbols**: (s) represents a solid; (l) a liquid, (g) a gas and (aq) an aqueous solution (a substance dissolved in water):

$$2HCl(aq) + Mg(s) \rightarrow MgCl_2(aq) + H_2(g)$$

   This balanced equation reads 'hydrogen chloride dissolved in water (hydrochloric acid) reacts with solid magnesium to form a solution of magnesium chloride in water and hydrogen gas'.

   Baking powder is often added when making cakes to make the cake 'rise'. One of the ingredients in baking powder is sodium hydrogencarbonate ($NaHCO_3$). The cake rises because carbon dioxide is given off:

   sodium hydrocarbonate → sodium carbonate + water + carbon dioxide
$$2NaHCO_3(s) \rightarrow Na_2CO_3(s) + H_2O\ (l) + CO_2(g)$$

5.5 The periodic table

Periods and groups

In the periodic table the elements are arranged in order of increasing proton (atomic) number. Elements with similar properties are placed in the same vertical group.

There are:
- seven horizontal rows of elements called **periods**
- eight vertical columns of elements called **groups**.

Fig. 5.7 The periodic table

1	2			transition elements								3	4	5	6	7	0
1 H Hydrogen 1																	**4** He Helium 2
7 Li Lithium 3	**9** Be Berylium 4											**11** B Boron 5	**12** C Carbon 6	**14** N Nitrogen 7	**16** O Oxygen 8	**19** F Fluorine 9	**20** Ne Neon 10
23 Na Sodium 11	**24** Mg Magnesium 12											**27** Al Aluminium 13	**28** Si Silicon 14	**31** P Phosphorus 15	**32** S Sulphur 16	**35.5** Cl Chlorine 17	**40** Ar Argon 18
39 K Potassium 19	**40** Ca Calcium 20	**45** Sc Scandium 21	**48** Ti Titanium 22	**51** V Vanadium 23	**52** Cr Chromium 24	**55** Mn Manganese 25	**56** Fe Iron 26	**59** Co Cobalt 27	**59** Ni Nickel 28	**64** Cu Copper 29	**65** Zn Zinc 30	**70** Ga Gallium 31	**73** Ge Germanium 32	**75** As Arsenic 33	**79** Se Selenium 34	**80** Br Bromine 35	**84** Kr Krypton 36
85.5 Rb Rubidium 37	**88** Sr Strontium 38	**89** Y Yttrium 39	**91** Zr Zirconium 40	**93** Nb Niobium 41	**96** Mo Molybdenum 42	**98** Tc Technetium 43	**101** Ru Ruthenium 44	**103** Rh Rhodium 45	**106** Pd Palladium 46	**108** Ag Silver 47	**112** Cd Cadmium 48	**115** In Indium 49	**119** Sn Tin 50	**122** Sb Antimony 51	**128** Te Tellurium 52	**127** I Iodine 53	**131** Xe Xeron 54
133 Cs Caesium 55	**137** Ba Barium 56	**139** La Lanthanum 57	**178.5** Hf Hafnium 72	**181** Ta Tantalum 73	**184** W Tungsten 74	**186** Re Rhenium 75	**190** Os Osmium 76	**192** Ir Iridium 77	**195** Pt Platinum 78	**197** Au Gold 79	**210** Hg Mercury 80	**204** Tl Thallium 81	**207** Pb Lead 82	**209** Bi Bismuth 83	**210** Po Polonium 84	**210** At Astatine 85	**222** Rn Radon 86
223 Fr Francium 87	**226** Ra Radium 88	**227** Ac Actinium 89	104	105	106	107	108	109									

KEY:

Relative atomic mass
Symbol
Name
Atomic Number

139 La Lanthanum 57	**140** Ce Cerium 58	**141** Pr Praseodymium 59	**144** Nd Neodymium 60	**147** Pm Promethium 61	**150** Sm Samarium 62	**152** Eu Europium 63	**157** Gd Gadolinium 64	**159** Tb Terbium 65	**162.5** Dy Dysprosium 66	**165** Ho Holmium 67	**167** Er Erbium 68	**169** Tm Thulium 69	**173** Yb Ytterbium 70	**175** Lu Lutetium 71
227 Ac Actinium 89	**232** Th Thorium 90	**231** Pa Procactinium 91	**238** U Uranium 92	**237** Np Neptunium 93	**242** Pu Plutonium 94	**243** Am Americium 95	**247** Cm Curium 96	**247** Bk Berkelium 97	**251** Cf Californium 98	**254** Es Einsteinium 99	**253** Fm Fermium 100	**256** Md Mendeleevium 101	**254** No Nobelium 102	**257** Lw Lawrencium 103

The main blocks of elements are in blue.

- The elements between the two parts of the main block are the **transition elements (metals)**.
- The bold stepped line of the table divides metals on the left-hand side from non-metals on the right-hand side.

Properties and reactions of group 1 elements (alkali metals)

EDEXCEL 360 C1a5

The first three elements in group 1 are lithium (Li), sodium (Na) and potassium (K). Because lithium, sodium and potassium belong to the same group in the periodic table, they must have very similar physical and chemical properties.

Element	Symbol	Appearance	Melting point in °C	Density in g/cm³
lithium	Li	soft, grey metal	181	0.54
sodium	Na	soft, light grey metal	98	0.97
potassium	K	very soft, blue/grey metal	63	0.86

Physical properties

- The alkali metals have low melting points, low boiling points and low densities compared with other metals.
- The melting points and boiling points of the alkali metals decrease on going down the group.
- They are good conductors of electricity.
- Their compounds are soluble in water

Chemical properties

- Lithium, sodium and potassium are very reactive metals. They are stored under oil to prevent them from reacting with air and water.
- The metals get more reactive going down the group.

Reaction with water

They are called **alkali metals** because they react with water to form alkaline solutions.

$$\text{lithium} + \text{water} \rightarrow \text{lithium hydroxide} + \text{hydrogen}$$
$$2Li(s) + 2H_2O(l) \rightarrow 2LiOH(aq) + H_2(g)$$
$$\text{sodium} + \text{water} \rightarrow \text{sodium hydroxide} + \text{hydrogen}$$
$$2Na(s) + 2H_2O(l) \rightarrow 2NaOH(aq) + H_2(g)$$
$$\text{potassium} + \text{water} \rightarrow \text{potassium hydroxide} + \text{hydrogen}$$
$$2K(s) + 2H_2O(l) \rightarrow 2KOH(aq) + H_2(g)$$

Reaction with oxygen

They burn in air or oxygen to form oxides, e.g.

$$\text{lithium} + \text{oxygen} \rightarrow \text{lithium oxide}$$
$$4Li(s) + O_2(g) \rightarrow 2Li_2O(s)$$

Each alkali burns with a characteristic flame colour: lithium – red; sodium – yellow; potassium – lilac.

The oxides dissolve in water to form alkalis, e.g.

$$\text{sodium oxide} + \text{water} \rightarrow \text{sodium hydroxide}$$
$$Na_2O(s) + H_2O(l) \rightarrow 2NaOH(aq)$$

Reaction with chlorine

They burn in chlorine to form a white solid of the metal chloride.

$$\text{potassium} + \text{chlorine} \rightarrow \text{potassium chloride}$$
$$2K(s) + Cl_2(g) \rightarrow 2KCl(s)$$

> **You should be able to predict the properties of rubidium knowing the properties of lithium, sodium and potassium.**

> **KEY POINT** These metals are in the same group of the periodic table because their reactions are very similar.

The next three members of group 1 are rubidium (Rb), caesium (Cs) and francium (Fr). Francium is the most reactive metal.

Properties and reactions of group 7 elements (the halogens)

EDEXCEL 360 **C1a5**

The first four elements in group 7 are fluorine (F), chlorine (Cl), bromine (Br) and iodine (I). They exist as diatomic molecules. Because fluorine, chlorine, bromine and iodine belong to the same group in the periodic table, they must have very similar chemical properties.

> **Remember diatomic means two atoms per molecule, e.g. Cl_2.**

Physical properties

Element	Symbol	Appearance	Melting point
fluorine	F	pale yellow gas	−220 °C
chlorine	Cl	yellow/green gas	−101 °C
bromine	Br	red/brown volatile liquid	−7 °C
iodine	I	dark grey crystalline solid, purple vapour when heated	114 °C

- The halogens have low melting points and low boiling points.
- The melting points and boiling points of the halogens increase going down the group.
- They are poor conductors of electricity.
- Apart from iodine, they are soluble in water and they react with water (see below).
- They are soluble in organic solvents such as hexane to form solutions with characteristic colours: chlorine – colourless; bromine – orange; iodine – purple.
- Many of their compounds are soluble in water.

Chemical properties
- Fluorine, chlorine, bromine and iodine are reactive non-metals.
- The non-metals get less reactive going down the group.
- Fluorine is the most reactive non-metal.

Reaction with water
When they are added to water, a mixture of acids is formed. The acid containing oxygen is a bleaching agent, e.g.

$$chlorine + water \rightarrow hydrochloric\ acid + chloric(I)\ acid$$
$$Cl_2(g) + H_2O(l) \rightarrow HCl(aq) + HClO(aq)$$

The test for chlorine gas is that it turns moist blue litmus red and then bleaches the litmus.

Reaction with metals

The halogens react with metals to form salts (halides). The name **halogen** means '**salt producer**'.

- **Chlorine** produces **chloride**s, e.g.

$$\text{iron} + \text{chlorine} \rightarrow \text{iron(III) chloride}$$
$$2Fe(s) + 3Cl_2(g) \rightarrow 2FeCl_3(s)$$

- **Bromine** produces **bromide**s, e.g.

$$\text{magnesium} + \text{bromine} \rightarrow \text{magnesium bromide}$$
$$Mg(s) + Br_2(l) \rightarrow MgBr_2(s)$$

- **Iodine** produces **iodide**s, e.g.

$$\text{sodium} + \text{iodine} \rightarrow \text{sodium iodide}$$
$$2Na(s) + I_2(s) \rightarrow 2NaI(s)$$

Displacement reactions of the halogens

> **KEY POINT**
> A more reactive halogen will displace a less reactive halogen from one of its compounds.

When chlorine is bubbled into an aqueous solution of potassium bromide, chlorine displaces the bromine. The colourless solution turns orange.

$$\text{chlorine} + \text{potassium bromide} \rightarrow \text{bromine} + \text{potassium chloride}$$
$$Cl_2(g) + 2KBr(aq) \rightarrow Br_2(aq) + 2KCl(aq)$$

Similarly

$$\text{chlorine} + \text{potassium iodide} \rightarrow \text{iodine} + \text{potassium chloride}$$
$$Cl_2(g) + 2KI(aq) \rightarrow I_2(aq) + 2KCl(aq)$$

$$\text{bromine} + \text{potassium iodide} \rightarrow \text{iodine} + \text{potassium bromide}$$
$$Br_2(l) + 2KI(aq) \rightarrow I_2(aq) + 2KBr(aq)$$

But when iodine is added to potassium chloride or potassium bromide there is *no* reaction.

There is also *no* reaction when bromine is added to potassium chloride.

Transition metals

Vanadium, chromium, manganese, iron, nickel and copper are examples of transition metals. They have the following general properties:

- high melting points, high boiling points and high densities
- a shiny appearance
- good conductors of heat and electricity
- form more than one positive ion, e.g. Fe^{2+} and Fe^{3+}, Cu^+ and Cu^{2+}
- compounds are often coloured, e.g. iron(II) sulphate is pale green and copper(II) sulphate is blue
- the transition metals and transition metal compounds are good catalysts, e.g. iron is used in the manufacture of ammonia and vanadium(V) oxide in the manufacture of sulphuric acid.

HOW SCIENCE WORKS

Food additives and packaging

Substances are often added to food (food additives) to enhance the colour, texture or flavour of the food, or to preserve it for longer. You may have noticed E-numbers on food labels. 'E' in front of the number means that the additive has been passed for food use in the European Community.

E numbers

E100–199	colourings
E200–299	preservatives
E300–399	acids, antioxidants and mineral salts
E400–599	vegetable gums, emulsifiers, stabilisers and anti-caking agents
E600 and above	flavour enhancers
E900–1500	miscellaneous

Food manufacturers have to state the food additives they have used by name or E number on the label as some people simply prefer to avoid eating food with additives, while others have mild to severe reactions to the additives.

Emulsifiers are added to emulsions to make them last longer. Examples of emulsions include milk, butter and salad dressing. Proteins such as those found in egg yolks are natural emulsifiers. Artificial emulsifiers are often related to natural emulsifiers; a common artificial emulsifier is E322 (lecithin – found in egg yolk).

Emulsifiers don't always stop the liquids separating for long enough so stabilisers may also be added. Many stabilisers are derived from seaweed or gums. Emulsifiers and stabilisers are often used to improve the texture of food.

Antioxidants are also commonly added to foods as preservatives. They help to stop the food from reacting with oxygen. Vinegar, honey and the juice from citrus fruits (containing vitamin C) are examples of natural antioxidants. Vitamin C, also called ascorbic acid or E300, is one of the most widely used antioxidants. Some modern preservatives are effective antioxidants. Some scientists believe they have evidence that corpses take longer to decay as a result of the increase in the amount of preservatives we eat in our food.

Additives are also used in other products, such as paint. Paint is an example of a water-based emulsion. The molecules are so small and evenly dispersed throughout the paint that it is called a colloid (the molecules are suspended, not dissolved in the water). A simple way to tell the difference between a solution and a colloid is to shine a beam of light through the liquid. If you can see the beam it is a colloid.

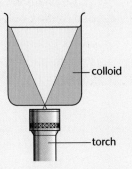

colloid

torch

HOW SCIENCE WORKS

Cooking food causes chemical changes in the food, normally involving proteins and/or cellulose, which can alter the appearance, taste and texture of the food.

The protein found in raw egg white, albumin, exists with weak bonds between the protein molecules. Cooking the egg causes the albumin to permanently change its shape (denatures) and allows it to form stronger bonds, which has the visible effect of stiffening the egg white and changing its colour from clear to white. The proteins in meat undergo similar changes during cooking, also changing the appearance and texture of the meat.

The cell walls in vegetables such as potatoes are made of insoluble cellulose molecules. When potatoes are heated, the cell walls break down making the potato easier to digest.

Changes in lifestyle are increasing the demands made on food manufacturers to provide food with clearer information labels, more convenience, fewer additives and a longer shelf life. 'Active' and 'intelligent' packagings have been in development for some time and in some cases have reached supermarket shelves already.

Active packaging is designed to change the condition of the food contained within it. It can absorb or release substances to maintain the freshness of the food, removing the need to add preservatives to the food itself. Active packaging can also change the temperature of the food, for example self-heating packages for soup or self-cooling packages for fizzy drinks.

Intelligent packaging is designed to indicate the condition of the food within it or the environment to which it has been exposed. Temperature 'spots' on packaging change colour if the package has been exposed to inappropriate temperatures that may affect the quality and safety of the food within it ('thermochromic' packaging). Similarly, other intelligent packages can indicate the presence of microbes, or exposure to light and physical shock.

Fresh   Still fresh – Consume immediately   No longer fresh

When combined, intelligent and active packaging could lead to less food wastage although there are concerns about whether it will be possible to recycle these more complex types of packaging. In turn, this is leading to more research into nature's answer to packaging, for example banana skins and egg shells, to develop bio-degradable packaging that mimics and enhances the protection they offer.

Exam practice questions

1. What is the formula of water? [1]
   (a) HO
   (b) HO_2
   (c) H_2O
   (d) H_2O_2

2. A compound has the formula $Cu(OH)_2$. What can we deduce from this formula? [1]
   (a) It contains five elements
   (b) It contains three elements
   (c) It contains covalent bonds
   (d) The ratio of copper, oxygen, hydrogen is 1:1:2

3. Which substance is a compound? [1]
   (a) air
   (b) milk
   (c) sea water
   (d) sugar

4. Finish the passage using words from the list.

   100 °C   compound   freezes   hydrogen   hydroxide
   potassium   isotope   neutrons   solvent   protons

   Water is a colourless liquid. It _____ at 0°C and boils at _____. It is a _____ of _____ and oxygen. Because water dissolves nearly every substance, it is known as the universal _____. Heavy water is deuterium oxide. Deuterium is an _____ of hydrogen. Its nucleus contains the same number of _____ but a different number of _____.

   When potassium is added to water, potassium _____ solution and hydrogen gas are formed. Most _____ compounds are soluble in water. [10]

5. What is meant by
   (a) an atom
   (b) a molecule
   (c) a group in the periodic table [3]

Exam practice questions

6   Answer questions (a) to (e) using only substances from the following list:

air   brass   calcium carbonate   copper   sodium chloride   sugar   water

   (a)   An element   [1]

   (b)   A mixture containing both elements and compounds   [1]

   (c)   A mixture of elements   [1]

   (d)   *Two* compounds containing only two elements   [2]

   (e)   *Three* compounds containing oxygen   [3]

   (f)   *Two* compounds that contain carbon   [2]

7.   Lettsium (Le) is an imaginary element. It has been given the following properties:

   (i)   it gives a blue colour with the flame test

   (ii)   it is a very good conductor of electricity

   (iii) it is soft and can be cut very easily with a knife

   (iv) it reacts violently with water, giving hydrogen and an alkaline solution

   (v)   it reacts with chlorine to form lettsium chloride.

   (a)   Divide the properties of lettsium into physical properties and chemical
            properties.   [1]

   (b)   Is lettsium a metal or a non-metal? Give a reason for your answer.   [2]

   (c)   Suggest how lettsium would be stored in the laboratory.   [1]

   (d)   **(i)**   In which group of the periodic table would you expect to find lettsium?   [1]

            (ii) Name *two* elements similar to lettsium.   [3]

   (e)   **(i)**   What would be the colour of lettsium chloride?

            (ii) What ions would be present in lettsium chloride?   [3]

6 Non-metal chemistry

The following topics are covered in this chapter:

- **Non-metals**
- **Limestone, cement and concrete**
- **Crude oil**
- **Organic compounds of carbon**
- **Polymers**
- **The air**
- **Nitrogen and its compounds**

6.1 Non-metals

Carbon

OCR A	C1.1
OCR B	C1g
AQA	C1.12.3
EDEXCEL 360	C1a6

Non-metals are found on the right-hand side of the diagonal in the periodic table (see Fig. 5.7).

Carbon forms both **inorganic** compounds and **organic** compounds. Inorganic chemistry is defined as 'the branch of chemistry concerned with compounds of elements other than carbon'. However, compounds such as carbon monoxide (CO), carbon dioxide (CO_2), calcium carbonate ($CaCO_3$) and sodium hydrogencarbonate ($NaHCO_3$) are considered to be inorganic compounds. Carbon is the sixth most abundant element in the Universe.

Where found	How found
atmosphere	carbon monoxide, carbon dioxide and methane
Earth's crust	as the pure element in diamond and graphite and as a mixture of the element or its compounds in coal, oil and natural gas (fossil fuels)
biosphere	as carbohydrates, proteins and fats
hydrosphere	as carbonates, hydrogencarbonates

Carbon dioxide

OCR A	C1.1
OCR B	C1a
AQA	C1 12.1
EDEXCEL 360	C1d6

If substances decompose when heated, the reaction is known as thermal decomposition.

About 0.04% of air is **carbon dioxide** (CO_2). It can be made by:
- burning carbon in air

$$\text{carbon} + \text{oxygen} \rightarrow \text{carbon dioxide}$$
$$C(s) + O_2(g) \rightarrow CO_2(g)$$

- reacting an acid with a carbonate

$$\text{copper(II)} + \text{hydrochloric} \rightarrow \text{copper(II)} + \text{water} + \text{carbon dioxide}$$
$$\text{carbonate} \qquad \text{acid} \qquad \text{chloride}$$
$$CuCO_3(s) + 2HCl(aq) \rightarrow CuCl_2(aq) + H_2O(l) + CO_2(g)$$

Sodium hydrogencarbonate is commonly known as bicarbonate of soda.

- heating carbonates and hydrogencarbonates

    calcium carbonate → calcium oxide + carbon dioxide
    $$CaCO_3(s) \rightarrow CaO(s) + CO_2$$
    sodium hydrogencarbonate → sodium carbonate + water + carbon dioxide
    $$2NaHCO_3(s) \rightarrow Na_2CO_3(s) + H_2O(l) + CO_2(g)$$

- burning hydrocarbons such as methane.

    methane + oxygen → carbon dioxide + water
    $$CH_4(g) + 2O_2(g) \rightarrow CO_2(g) + 2H_2O(g)$$

- the process of respiration

    glucose + oxygen → carbon dioxide + water
    $$C_6H_{12}O_6 + 6O_2(g) \rightarrow 6CO_2(g) + 6H_2O(g)$$

KEY POINT

The test for carbon dioxide is to bubble the gas into limewater (calcium hydroxide solution). A white, cloudy compound (calcium carbonate) is formed. The limewater is said to turn milky.

$$Ca(OH)_2(aq) + CO_2(g) \rightarrow CaCO_3(s) + H_2O(l)$$

The carbon cycle

OCR A   P2.4
OCR B   C2f

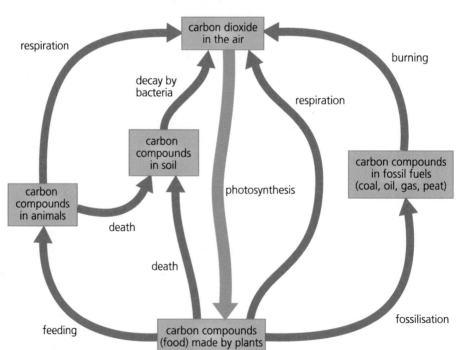

Fig. 6.1 The carbon cycle

The three main processes in the carbon cycle are:
- respiration
- burning (combustion)
- photosynthesis

The level of carbon dioxide in the atmosphere is gradually increasing, resulting in **global warming**.

6.2 Limestone and concrete

Inorganic compounds of carbon

OCR B   C2b
AQA A   C1.12.1
EDEXCEL 360

Limestone and **marble** are both forms of **calcium carbonate** ($CaCO_3$). Limestone is quarried and used in the manufacture of:

● lime (calcium oxide)
● cement
● iron.

Removing large amounts of limestone from the ground:
● destroys the landscape
● takes up valuable land-space
● creates noise and dust pollution
● increases traffic in the quarry vicinity.

Lime

> Lime is also known as quicklime.

Lime (calcium oxide) is made by the thermal decomposition of **limestone**. Lumps of limestone are heated strongly in a limekiln:

$$\text{calcium carbonate} \rightarrow \text{calcium oxide} + \text{carbon dioxide}$$
$$CaCO_3(s) \rightarrow CaO(s) + CO_2(g)$$

Most of the manufactured lime is reacted with water (a process known as **slaking**) to form calcium hydroxide (slaked lime).

Lime and slaked lime are used to:
● reduce the acidity of soil
● remove acidic gases such as sulphur dioxide and carbon dioxide in factory chimneys.

Concrete

Concrete is made when cement, sand and water are mixed together and allowed to set. During the process of setting, calcium hydroxide reacts with carbon dioxide from the air to form calcium carbonate:

> For OCR B only.

$$\text{calcium hydroxide} + \text{carbon dioxide} \rightarrow \text{calcium carbonate} + \text{water}$$
$$Ca(OH)_2(aq) + CO_2(g) \rightarrow CaCO_3(s) + H_2O(l)$$

Sometimes concrete is reinforced by allowing the concrete to set around steel supports. This brings together the properties of the hardness of concrete and the flexibility and strength of steel. When a load is placed on a reinforced concrete beam, it bends slightly. If a concrete beam were used, it might break.

The above reactions can be summed up in the 'calcium cycle'.

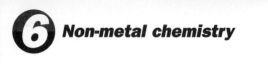

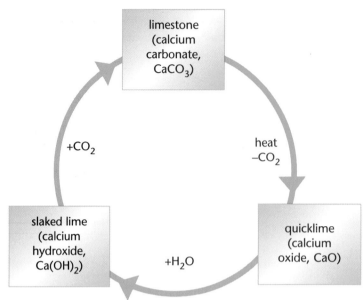

Fig. 6.2 The calcium cycle.

6.3 Crude oil

Fractional distillation

OCR A   C2.2
OCR B   C1d
AQA   C1.12.3
EDEXCEL 360   C1b7

See pages 95–6 for more about alkanes.

Crude oil is an ancient **biomass** found in rocks. Crude oil, coal and natural gas are **fossil fuels**. Fossil fuels are **non-renewable** fuels. Non-renewable fuels take a very long time to make and they are used up faster than they can be formed.

Crude oil is a **mixture** of a very large number of compounds. Most of the compounds in crude oil consist of **hydrocarbons** of varying lengths. Hydrocarbons are compounds made from hydrogen and carbon atoms *only*. Most of the hydrocarbons are **alkanes**.

The hydrocarbons in crude oil are separated in **fractions** by evaporating the crude oil and allowing it to condense at a number of different temperatures. A fraction is a mixture of substances (in this case, hydrocarbons) with similar boiling points. This process of separation is called **fractional distillation**. Figure 6.3 shows how the different fractions can be obtained from crude oil.

During boiling, the intermolecular forces between the molecules are broken. These forces are stronger between larger hydrocarbon molecules than between smaller hydrocarbon molecules. The larger hydrocarbons have higher boiling points (e.g. lubricating oil) and 'exit' from the bottom of the fractionating column. The smaller hydrocarbons (e.g. petrol) have lower boiling points and 'exit' from the top of the fractionating column.

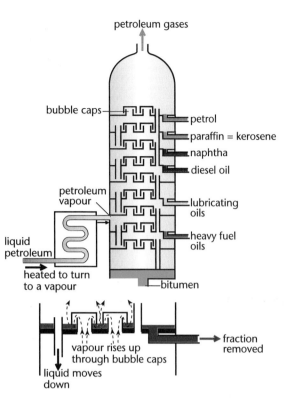

Fig. 6.3 Separation of fractions

The table gives some information about the fractions and their uses.

Fraction	Boiling range	Number of carbon atoms per molecule	Use
petroleum gas	below 40°C	1–4	fuel for cooking and heating
petrol	40–75°C	5–10	fuel for cars
naphtha	75–160°C	7–14	for manufacture of plastics and detergents
paraffin	160–250°C	11–16	fuel for aircraft
diesel oil	250–300°C	16–20	fuel for cars and trains
lubricating oil	300–350°C	20–35	for motor engines
heavy fuel oils	350–400°C	30–70	for ships and power stations
bitumen	above 400°C	more than 70	road tar

6.4 Organic compounds of carbon

Alkanes

OCR B   C1e
AQA   C1.12.3

Organic chemistry is 'the branch of chemistry concerned with compounds of carbon' such as methane (CH_4), ethene (C_2H_4), ethanol (C_2H_5OH), ethanoic acid (CH_3COOH) and ethyl ethanoate ($CH_3COOC_2H_5$).

Alkanes are saturated hydrocarbons. This means they contain only single carbon–carbon **covalent** bonds.

Name	Formula	Structure	State at room temp	Melting point	Boiling point
methane	CH_4	H | H–C–H | H	gas		
ethane	C_2H_6	H H | | H–C–C–H | | H H	gas	Increases down the family	Increases down the family
propane	C_3H_8	H H H | | | H–C–C–C–H | | | H H H	gas		
butane	C_4H_{10}	H H H H | | | | H–C–C–C–C–H | | | | H H H H	gas		
pentane	C_5H_{12}	H H H H H | | | | | H–C–C–C–C–C–H | | | | | H H H H H	liquid		

Alkanes

● have a general formula C_nH_{2n+2}. Thus $C_{20}H_{42}$ is an alkane
● have very few reactions apart from burning in air or oxygen.

Burning alkanes

When alkanes burn in **excess** air or oxygen, **carbon dioxide** and **water vapour** are formed:

methane + oxygen → carbon dioxide + water
$$CH_4(g) + 2O_2(g) \rightarrow CO_2(g) + 2H_2O(g)$$

When alkanes burn in **limited** air, **carbon monoxide** and **water vapour** are formed:

methane + oxygen → carbon monoxide + water
$$2CH_4(g) + 3O_2(g) \rightarrow 2CO(g) + 4H_2O(g)$$

Carbon monoxide is very poisonous.

Alkenes

OCR B   C1e
AQA   C1.12.4

Alkenes are **unsaturated** hydrocarbon molecules. They contain carbon–carbon **double** bonds.

Name	Formula	Structure	State at room temp	Melting point	Boiling point
ethene	C_2H_4		gas	Increases down the family	Increases down the family
propene	C_3H_6		gas		

Alkenes

- have a general formula C_nH_{2n}. Thus $C_{20}H_{40}$ is an alkene
- are much more reactive than alkanes.

Reactions of alkenes

- Alkenes burn in excess air or oxygen to form carbon dioxide and water vapour:

ethene + oxygen → carbon dioxide + water

$C_2H_4(g) + 3O_2(g) → 2CO_2(g) + 2H_2O(g)$

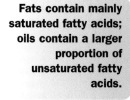

You must state the exact colour change. It is wrong to say that the solution turns clear. All solutions are clear.

- When alkenes are bubbled through a solution of bromine, the solution changes from **red-brown** to **colourless**:

propene + bromine → 1,2-dibromopropane

$C_3H_6(g) + Br_2(aq) → C_3H_6Br_2(l)$

This is an addition reaction.

> **KEY POINT**
> An addition reaction is a reaction in which two or more molecules combine to form a single product.

Other addition reactions:

- Alkenes react with steam at high temperatures and high pressures in the presence of a catalyst to form alcohols:

ethene + steam → ethanol

$C_2H_4(g) + H_2O(g) → C_2H_5OH(l)$

This reaction is used to manufacture ethanol.

- When passed with hydrogen over a nickel catalyst at 200°C, **alkanes** are formed:

alkene + hydrogen → alkane

$C_nH_{2n} + H_2 → C_2H_{2n+2}$

Fats contain mainly saturated fatty acids; oils contain a larger proportion of unsaturated fatty acids.

This type of reaction is used to make margarine from vegetable oils. Vegetable oils are highly unsaturated liquids. By reacting with hydrogen, the unsaturated oils form saturated oils, which are solids.

Cracking

> **KEY POINT**
>
> Cracking is the breaking down of a long-chain hydrocarbon into smaller molecules. The long-chain hydrocarbons are passed over a catalyst at a high temperature. The products of cracking include alkanes and alkenes and sometimes hydrogen.

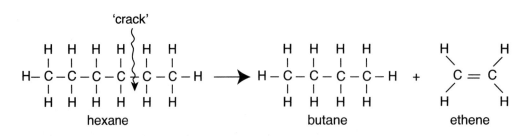

hexane   butane   ethene

Industry can meet its demands for petrol (mainly octane C_8H_{18}) and other smaller molecules by cracking petroleum fractions, such as lubricating oil ($C_{30}H_{62}$). This process can be demonstrated in the laboratory using the apparatus shown in Fig. 6.4.

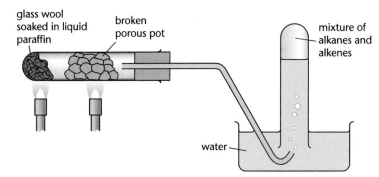

Fig. 6.4 Apparatus for cracking petroleum fractions in the laboratory.

Most countries depend upon oil from a few oil-producing nations. Oil is transferred either in tankers or piped over large distances to the country of destination. The oil-producing countries can dictate the price of oil. Oil pipes can be sabotaged. Sometimes tankers get damaged, spilling oil into the sea and causing enormous pollution problems.

6.5 *Common organic compounds*

Polymers

OCR A	C2.2
OCR B	C1e
AQA B	C1.12.4
EDEXCEL 360	C1b8

Alkenes can be used to make polymers such as poly(ethene) and poly(propene). In these reactions, many small molecules (monomers) join together to form very large molecules (polymers).

> **KEY POINT**
>
> The process of joining together a large number of small molecules (monomer) to form a large molecule (polymer) is called polymerisation.

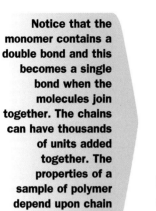

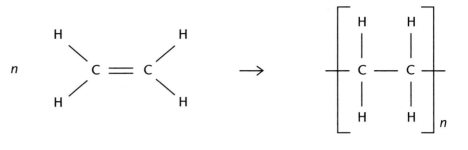

Notice that the monomer contains a double bond and this becomes a single bond when the molecules join together. The chains can have thousands of units added together. The properties of a sample of polymer depend upon chain length.

In this reaction the monomer is ethene and the polymer poly(ethene) commonly known as polythene.

Polymers have properties that depend upon what they are made from and the conditions under which they are made. You may have made **slime** with different viscosities from the polymer poly(ethenol). The viscosity depends on how the polymer molecules are cross-linked. The more linkages there are, the greater will be the viscosity.

You should be able to work out that the name of the monomer from which poly(ethenol) is made is ethenol.

Addition polymers such as **poly(ethene)** and **poly(vinyl chloride)** (PVC) have many uses.

They have replaced traditional materials such as metals, paper, cardboard and rubber.

The table lists common polymers and their uses.

Polymer (alternative name, if any)	Monomer (alternative name, if any)	Uses
poly(ethene) (polythene)	ethene	cling film, plastic bags, coating for electric wire
poly(propene) (PP)	propene	food packaging, car parts
poly(chloroethene) poly(vinyl chloride) (PVC)	chloroethene (vinyl chloride)	plastic raincoats, shower curtains, insulation for electrical wiring
poly(styrene)	styrene	disposable cups, packaging, egg boxes, CD 'jewel' case
poly(tetrafluoroethene) (Teflon)	tetrafluoroethene	coating non-stick pans, making Gore-Tex
poly(methyl methacrylate) (Perspex)	methyl methacrylate	windscreens, aircraft windows, protective glass in hockey stadiums and large aquariums

Make sure that you can name the polymer from monomers and the monomer from polymers.

Polymers have many useful applications. New uses are being developed, for example: new packaging materials, waterproof coatings for fabrics, dental polymers, wound dressings, hydrogels and smart materials, including shape memory polymers.

Many polymers are not biodegradable, so they are not broken down by micro-organisms. This can lead to problems with waste disposal.

All addition polymers are **thermoplastics**. This means that they melt when heated and they can then be remoulded into different shapes. Thermosetting plastics (e.g. **Bakelite**) do not melt on heating; instead they decompose. Many plastics are **non-biodegradable**.

> **KEY POINT** A non-biodegradable plastic cannot be decomposed by bacteria in the soil.

Disposal of plastics

Plastics with this sign can be recycled.

- Recycle – when you dispose of your rubbish you may be asked to separate your recyclable plastic from non-recyclable plastics.
- Burn in plentiful supply of air. Plastics made up of carbon and hydrogen burn to give carbon dioxide and water. The heat energy can be used to supply power to factories. Plastics such as Teflon and PVC give off poisonous gases when they are burnt, causing air pollution.
- Landfill – this gets the plastics out of sight, but, since they do not rot away they cause land pollution.

Alcohols

OCR B   C1c
AQA   C1.12.4
EDEXCEL 360   C1b8

Alcohols are organic compounds that contain the –OH group. The substance we call alcohol in everyday life is **ethanol** (C_2H_5OH). Ethanol is present in various concentrations in a large number of drinks such as beer, lager, gin and whisky. It can be made by either a **batch** method or a **continuous process** method. In the fermentation process sugars are dissolved in water and yeast added. The mixture is left at room temperature in the absence of air. At the end of the fermentation, the mixture is fractionally distilled to obtain the ethanol:

$$glucose \rightarrow ethanol + carbon\ dioxide$$
$$C_6H_{12}O_6(aq) \rightarrow 2C_2H_5OH(aq) + 2CO_2(g)$$

In the continuous process, ethene is reacted at high temperature and pressure with steam in the presence of a catalyst of phosphoric acid:

$$ethene + water \rightarrow ethanol$$
$$C_2H_4(g) + H_2O(g) \rightarrow C_2H_5OH(g)$$

Remember, alcohol is a drug. Excess can cause:
- loss of inhibitions (you may do something you regret doing)
- blurred vision, slurred speech and loss of balance
- liver failure
- high blood pressure
- certain types of cancer.

The table compares the two processes.

	Fermentation	Batch
type of process	batch	continuous
purity of product	needs to be filtered and distilled	pure (ethanol is the only product)
resources	renewable (sugar from sugar cane)	non-renewable (ethene is obtained by cracking crude oil)
rate of process	slow	fast

Ethanol could be the fuel of the future. There are flex-fuel engines that are powered by ethanol or gasoline or any combination of both. They go under a variety of names such as Ethanol Fuels, Biofuel, Biodiesel and Alternative Fuels. The need to develop these fuels is to reduce our dependency on oil. Maybe in the future we will be fermenting our own fuel.

Solvent + solute = solution.

Ethanol is also a solvent, and is used to make perfumes and cosmetics.

> **KEY POINT**
>
> A solvent is a liquid that dissolves other chemicals (solutes) to form a solution.

Acids and esters

Carboxylic acids

If ethanol is left in air, it slowly oxidises to form ethanoic acid:

$$\text{ethanol} + \text{oxygen} \rightarrow \text{ethanoic acid} + \text{water}$$
$$C_2H_5OH(l) + 2[O] \rightarrow CH_3COOH(l) + H_2O(l)$$

> **[O] is used to represent oxidation.**

Vinegar is a solution of **ethanoic acid** in water. Ethanoic acid has a sharp, bitter taste. You may have tasted **citric acid** in oranges and lemons (citrus fruits) and **lactic acid** in sour milk.

Organic acids have the same properties as inorganic acids such as hydrochloric acid and sulphuric acid except the reactions are much slower.

Esters

Alcohols react with carboxylic acids to form an ester and water. Ethanol reacts with ethanoic acid to give ethyl ethanoate. Esters have pleasant smells.

Ester	Smell
ethyl ethanoate	pear drops (nail varnish remover)
ethyl methanoate	raspberry
ethyl butanoate	pineapple
pentyl ethanoate	pears

Perfumes are esters that are either man-made (synthesised) or extracted from natural materials such as lavender, roses and lily-of-the-valley.

Perfumes have the following properties.

> **Ethyl ethanoate is an organic liquid and therefore can act as a solvent for organic substances such as nail varnish.**

Property	Reason
evaporates easily	so that the perfume particles can easily reach the nose
insoluble in water	prevents it from being easily washed away
non-toxic	prevents poisoning the wearer
does not irritate the skin	so it is not uncomfortable to wear
does not react with water	so it does not react with perspiration

6.6 The air

How the atmosphere has changed

The air consists of a mixture of gases. The main gases in the air are shown in the table.

Gas	Element or Compound	Percentage by volume
nitrogen	element	78
oxygen	element	21
argon	element	0.9
carbon dioxide	compound	0.035
water vapour	compound	0.5–5

The process by which the Earth's atmosphere evolved is not really understood. One of the theories is that about 4 million years ago the atmosphere consisted of carbon dioxide (CO_2), ammonia (NH_3) and a small amount of methane (CH_4). These gases had been released during volcanic eruptions. Ammonia reacted with the rocks to form nitrogen (an unreactive gas) and water.

The compounds contained the elements essential to life: carbon, hydrogen, nitrogen and oxygen. Carbon can form long, complex chain molecules. During the frequent thunderstorms, the energy from the lightning caused the elements to combine to form amino acids. Slowly, photosynthetic organisms developed and this increased the percentage of oxygen and decreased the amount of carbon dioxide. Some of the carbon dioxide dissolved in the oceans and seas.

> **Nitrogen is unreactive so very little was removed once photosynthesis started. The percentage of nitrogen thus slowly increased.**

Air pollution

> **KEY POINT**   Air pollution is air that has chemicals in high enough concentrations to harm living things or to damage non-living things.

The table lists the major sources of air pollutants and how they arise.

Polluting substance	Main source	Effects on the environment
particulates	burning diesel fuel, soot, dust, smoke	lung problems
carbon dioxide	complete combustion of fossil fuels	greenhouse effect
carbon monoxide	incomplete combustion of fossil fuels	reduces oxygen carriage in the blood
methane	flatus of cattle, rice fields	greenhouse effect
oxides of nitrogen	car exhaust fumes	acid rain
sulphur dioxide	burning fossil fuels, volcanic eruptions	acid rain

The **sulphur** compounds present in **fossil fuels** are a very big source of **air pollution** in most industrial countries because sulphur dioxide is formed when fossil fuels are burned. If the sulphur is removed and collected it can be used as a supply of sulphur, and reduces pollution.

The main sulphur compound in natural gas (methane) is **hydrogen sulphide**. To remove this gas, natural gas is passed through specially prepared carbon powder. The hydrogen sulphide is adsorbed by the carbon. (You may have a cooking hood in your kitchen to remove smells. This works in a similar manner to the industrial removal of hydrogen sulphide.)

It is less easy to remove the sulphur directly from solid fuels such as coal. In large power plants the sulphur dioxide is removed from the waste gases after combustion. The gases are passed through scrubbers which contain calcium carbonate, calcium oxide and water. Sulphur dioxide and other acidic gases are removed. One of the products of this reaction is calcium sulphate, which is used in the manufacture of paints, ceramics and paper.

Catalytic converters can be used to remove carbon monoxide and oxides of nitrogen from exhaust fumes.

carbon monoxide + nitrogen(II) oxide → carbon dioxide + nitrogen

$$2CO(g) + 2NO(g) \rightarrow 2CO_2(g) + N_2(g)$$

Obtaining nitrogen and oxygen from the air

EDEXCEL 360 C1b7

Nitrogen, oxygen and argon are obtained from air by the fractional distillation of liquid air.

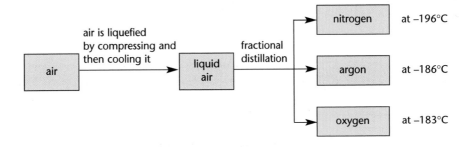

Fig. 6.5 Fractional distillation.

6.7 Nitrogen and its compounds

The nitrogen cycle

OCR A   C3.1

The diagram below summarises the nitrogen cycle. It describes how **nitrogen** and nitrogen-containing compounds in nature are interchanged.

The nitrogen fixing bacteria and pea plants have a mutualistic relationship. The bacteria are provided with some food from the plant and they fix nitrogen for the plant to use.

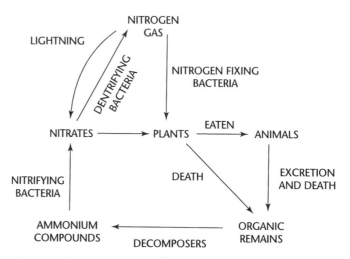

There are three main stages.

Nitrifying bacteria – these bacteria live in the soil and convert ammonium salts to nitrates. They need oxygen to do this.

Denitrifying bacteria – these bacteria in the soil are the enemy of farmers. They convert nitrates into nitrogen gas. They do not need oxygen.

Nitrogen fixing bacteria – these bacteria live in the soil or special bumps called nodules in the roots of plants from the pea and bean family (legumes). They convert nitrogen gas to nitrogen compounds.

Ammonia

OCR B   C2h
AQA   C1.12.3
EDEXCEL 360 C1a6

Manufacture of ammonia

Nitrogen is a non-metal that makes up nearly 80% of the atmosphere.

> **KEY POINT** Nitrogen combines with hydrogen to form ammonia (NH_3).

Ammonia is manufactured from nitrogen and hydrogen in a process known as the Haber process. Nitrogen is obtained from the air and hydrogen from methane. Figure 6.6 summarises the Haber process.

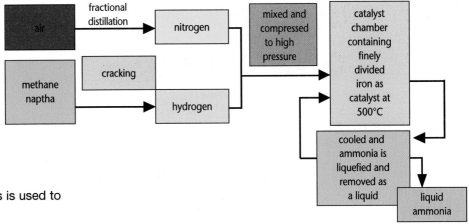

Fig. 6.6 The Haber process is used to manufacture ammonia.

The equation for the reaction is:

$$N_2(g) + 3H_2(g) \rightleftharpoons 2NH_3(g)$$

The reversible sign $\rightleftharpoons$ means that the products can decompose and re-form the reactants. By choosing the best conditions, chemists can produce the highest **yield** of ammonia in a reasonable length of time.

The conditions for the Haber process are:
- high pressure
- high enough temperature to give a reasonable rate of reaction
- iron catalyst.

Fertilisers

Useful elements in fertilisers

 KEY POINT **Fertilisers are chemicals that provide plants with essential chemical elements.**

The essential elements for a plant to grow are **nitrogen**, **phosphorus** and **potassium**. These essential elements can be obtained from either natural sources or artificial sources.

The table summarises the importance of these elements.

Element	Importance to growing plant	Natural sources	Artificial sources
nitrogen	for growth of stems and leaves	manure, bird droppings, dried blood	ammonium nitrate ammonium sulphate urea
phosphorus	for root growth	bone meal	ammonium phosphate
potassium	for flowers and fruit	wood ash	potassium sulphate

Fertilisers must be soluble in water. They are used:
- to replace elements used up by previous crops
- to add more nitrogen so plants increase in size.

Figure 6.7 summarises how the fertiliser ammonium nitrate is made.

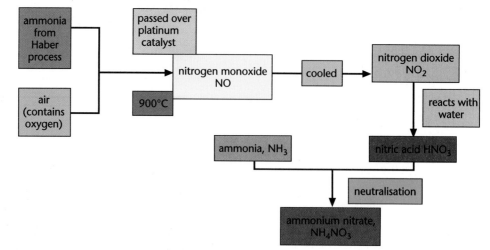

Fig. 6.7 The process for the manufacture of ammonium nitrate.

Overuse of fertilisers

Too much fertiliser can be harmful. When it rains, the fertilisers can be washed out of the soil and end up in ponds, lakes and rivers. This leads to **eutrophication**.

Eutrophication describes the following process:

> **Nitrates are dangerous to humans. They prevent haemoglobin in the blood from transporting oxygen around the body properly. Sewage escaping into ponds, lakes and rivers causes a similar problem.**

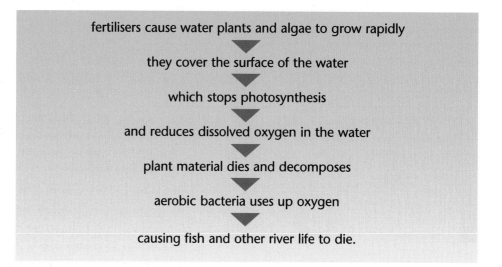

Yields of crops are reduced by pests and disease. Intensive farmers use pesticides such as **insecticides** (which kill insect pests, e.g. mosquitoes), **rodenticides** (for killing rodents such as rats and mice) and **herbicides** (for killing unwanted plants and weeds) to improve the quantity of their stocks.

Organic farmers have to use different methods such as:
- rotation of crops
- growing resistant crop varieties
- ensuring that plants are strong and healthy
- physical means such as grease bands on trees trunks and copper bands around the stems of plants
- biological pest controls such as nematodes
- sprays made from chemicals found in plants such as rotenone (obtained from tropical plants) and pyrethrum (obtained from foxgloves).

Farmers have to follow the UK national standards if they want to claim that their products are **organic**.

HOW SCIENCE WORKS

Scientific discoveries

Many important discoveries have been made by accident. Accidental discoveries are known as serendipity.

In 1928, a chemist failed to clear up his experiment. He left a mixture of an unsaturated organic compound and other chemicals in a test tube for two weeks. The compound polymerised to form what is now known as neoprene rubber. One of the chemists in the laboratory was Carothers. He studied the new substance and discovered its structure. He went on to use his research to make nylon. He is said to be the founder of polymer chemistry.

Poly(tetrafluoroethene) (Teflon) was accidentally discovered in 1937. The scientists were trying to make a new chlorofluorocarbon (CFC) refrigerant when unexpectedly fluoroethene polymerised in its container. Teflon is now used for non-stick pots and pans.

By the end of World War II, chemists were beginning to understand how to alter the properties of both natural and man-made polymers. During one such experiment a polymer got stuck to the equipment being used – superglue was discovered. In the 1970s, fumes from superglue were observed to condense around the fingers of a policeman. He had discovered a way of taking fingerprints.

Post-it® notes were invented by a chemist working for a company called 3M. He was trying to design a strong adhesive. Instead he developed an adhesive that was very weak. It was initially used as an adhesive for bookmarks, but the marketing department of 3M rejected them as useless. A secretary was told to get rid of them: instead she found a use for them. The rest is history.

Nowadays polymers are specifically designed and developed for specific purposes. A few examples are: new packaging materials, waterproof coatings for fabrics, dental polymers, wound dressings, hydrogels, smart materials and shape memory polymers.

HOW SCIENCE WORKS

Modern car components are made from polymers.

Component	Made from
bumpers	poly(acrylonitrilebutadienestyrene) (ABS)
headlights	polycarbonates
tyres	polyisobutene
	Kevlar

Kevlar was developed to make a 'super fibre'. It is five times as strong as steel on a weight for weight basis. It is lightweight, flexible and comfortable. Apart from tyres, one of its important uses is in bullet-proof clothing. It is amazing to think that this material is made up of only carbon, hydrogen, nitrogen and oxygen.

Gore-Tex® materials are made from an expanded polytetrafluoroethene (Teflon) and other fluoropolymers. It is used for outdoor clothing because it is waterproof, windproof and allows sweat to escape. Before its invention in 1978, no waterproof and breathable fabric existed. People got wet from the inside even if it wasn't raining.

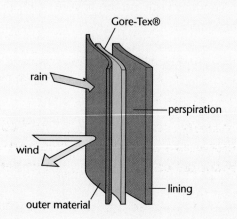

Gore-Tex® material is composed of a thin, porous fluoropolymer membrane bonded to other polymers such as nylon or polyesters. The membrane has billions of pores per square cm. Each pore is 20 000 times smaller than a water droplet. Liquid water cannot pass through the pores but water vapour can. The outer fabric is treated with water repellent.

Exam practice questions

1. Which reaction does *not* produce carbon dioxide? **[1]**
   (a) burning fossil fuels
   (b) fermentation
   (c) manufacture of lime
   (d) photosynthesis

2. Which statement is true about fertilisers? **[1]**
   (a) They must be soluble in water
   (b) They must contain nitrogen
   (c) They must contain phosphorus
   (d) They must contain potassium

3. What is the monomer of the isomer poly(isobutene)? **[1]**

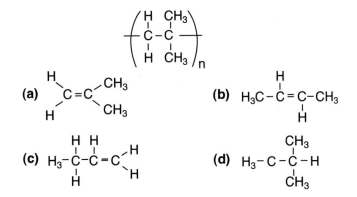

4. Finish the passage using words from the list.

   alkanes   alkenes   bromine   colourless   cracked   hydrogen   mixture
   monomers   petrol   polymers

   The chief source of organic compounds is the naturally occurring _____ called
   crude oil. Crude oil contains hydrocarbons. Hydrocarbons are compounds containing
   only carbon and _____. Saturated hydrocarbons are called _____. Unsaturated
   hydrocarbons are called _____.
   Unsaturated hydrocarbons can be distinguished from saturated hydrocarbons by adding
   _____ water. The solution changes from orange-brown to _____.
   Unsaturated hydrocarbons can combine together in large numbers to form _____.
   The small molecules are called _____.
   One of the fractions obtained during the fractional distillation of crude oil is
   lubricating oil. This can be _____ to form smaller molecules, such as _____. **[10]**

5. Choose from the following list of words, the gas described by each statement. A gas may
   be used once, more than once or not at all.

   carbon monoxide   carbon dioxide   methane   nitrogen
   oxides of nitrogen   sulphur dioxide

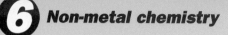

Exam practice questions

   (a)   The commonest gas in the atmosphere.   **[1]**

   (b)   The gas formed by incomplete combustion of a fuel.   **[1]**

   (c)   The gas formed by the bacterial decay of vegetable matter.   **[1]**

   (d)   The gas that causes acid rain.   **[1]**

   (e)   The gases formed by an internal combustion engine.   **[3]**

   (f)   The gases produced in the atmosphere by lightning.   **[1]**

   (g)   Two greenhouse gases.   **[2]**

6.   Fertilisers are added to the soil to help plants to grow.

   (a) Write down the names and formulae of three fertilisers that contain nitrogen.   **[3]**

   (b) Write down the name and formula of a fertiliser that does not contain nitrogen.   **[1]**

   (c)   **(i)**   Using too much fertiliser can kill fish. Below are five sentences describing how this happens, but they are in the wrong order. Rewrite them in the correct order.   **[3]**

            Algae grow well on the fertiliser and cover the river.

            Excess fertiliser dissolves in rain water and drains into rivers.

            There is little oxygen left for the fish and they die.

            The algae die and bacteria decompose them.

            The bacteria use up most of the oxygen in the water.   **[3]**

      (ii)   What name is given to the process described in **(c)(i)**?   **[1]**

7.   The flow chart for making ammonia by the Haber process is shown below.

   (a)   **(i)**   Write the balanced equation for the reaction between nitrogen and hydrogen to make ammonia.

      (ii)   Name the catalyst used in the process.   **[3]**

N_2 H_2 → gases are pressurised, heated and mixed → catalyst → mixture of gases is cooled → NH_3

      After passing the gases over the catalyst, the mixture of gases contains hydrogen, nitrogen and ammonia.

   (b)   Why are hydrogen and nitrogen present in the mixture?   **[1]**

      The mixture is cooled to −80 °C.

   (c)   Explain why only ammonia turns into a liquid at −80°C.   **[1]**

      The graph shows the yield of ammonia at different pressures and temperatures.

yield of ammonia in mole percent

70 — 350°C

60 —

50 —

40 — 450°C

30 —

20 — 550°C

10 —

0

100   200   300   400

pressure in atmospheres

Exam practice questions

(d) **(i)** Deduce from the graph the best conditions for obtaining a good yield of ammonia.

(ii) Why are these conditions *not* used in the manufacture of ammonia?

(iii) What pressure and temperature are used in the Haber process? **[5]**

8. C_8H_{18} is a hydrocarbon present in petrol.
   (a) **(i)** What is meant by a hydrocarbon?
   (ii) Is C_8H_{18} an alkane or an alkene? Explain your answer.
   (iii) Give *one* chemical test to distinguish between an alkane and an alkene. **[5]**

When petrol is burned in a motor car engine, the exhaust fumes contain approximately 10% carbon dioxide, 4% carbon monoxide, 4% oxygen, 2% hydrogen, 0.2% oxides of nitrogen and 0.2% hydrocarbons.

(b) The gases listed above only make up 20.4% of the exhaust gases.
What gas makes up the greater percentage of the remaining gases? **[1]**

(c) Why are the following gases present in the exhaust fumes?
   (i) carbon dioxide
   (ii) carbon monoxide
   (iii) oxides of nitrogen
   (iv) hydrocarbons

[4]

Hydrogen is present because some of the petrol is cracked to form an alkene and hydrogen.

(d) Write an equation for the cracking of C_8H_{18} to form hydrogen. **[2]**

One way of removing oxides of nitrogen and carbon monoxide is to pass the exhaust gases over a catalyst. The two gases react to form nitrogen and carbon dioxide.

(e) **(i)** Write the equation for the reaction that occurs when nitric oxide (N_2O) and carbon monoxide are passed over a catalyst in an exhaust pipe of a car.

(ii) Why is it better to let carbon dioxide (a greenhouse gas) escape into the air rather than carbon monoxide? **[3]**

9. An organic compound has the formula $C_2H_2Cl_2$.
   (a) Draw the structures of *two* unsaturated compounds that have this formula. **[2]**
   (b) Draw the structure, showing two units, of the polymer that can be obtained from one of these monomers. **[2]**
   (c) What name is given to this type of polymerisation? **[1]**
   One of the ways to dispose of polymers is to burn them in excess oxygen.
   (d) Suggest two gases that might be formed (other than water) when the polymer from **(b)** is burned in excess oxygen. **[2]**

7 mineral chemistry

The following topics are covered in this chapter:

- **The structure of the Earth**
- **Rocks as building materials**
- **Metals from rocks**
- **Alloys**
- **Salts**

7.1 The structure of the Earth

The Earth

OCR A	P1.1
OCR B	C2c
AQA	C1.12.6
EDEXCEL 360	C1a6

The Earth consists of:
- a **core** consisting mainly of nickel and iron
- a **mantle** that behaves like a solid but allows very slow convection currents to transfer energy from the centre to the surface
- a **crust** (**lithosphere**) whose thickness varies between 10 km (under oceans) and 65 km (under mountains). It is made up of a mixture of minerals. The most abundant elements in the crust are **silicon**, **oxygen** and **aluminium**.

> The Earth is a sphere which is very slightly flattened at the poles.

It was once thought that the features of the Earth's surface were the result of the shrinking of the crust as the Earth cooled down following its formation. This theory has been replaced by the **tectonic plate theory**.

Plate tectonics

The Earth's crust and the upper part of the mantle are cracked into a number of large pieces of thin rigid plates, called **tectonic plates**. The plates are found on top of the mantle because they are less dense than the mantle.

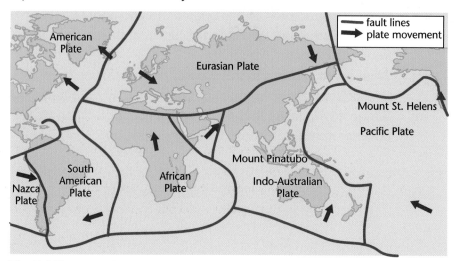

Fig. 7.1 The tectonic plates.

- **oceanic plates** under oceans
- **continental plates** forming continents.

Oceanic plates are denser than continental plates.

This is about the same rate as your nails grow.

There are **convection currents** within the Earth's mantle. These are driven by heat released by natural radioactive processes. The currents cause the plates to move at relative speeds of between 1 to 10 centimetres per year. The plates are always in motion. The movements can be sudden and disastrous. At the plate boundaries, earthquakes and volcanic eruptions occur. Sometimes mountains are formed and sometimes oceanic trenches.

Types of tectonic plate movement and their effects

Plates move in three different ways. They may
- slide past each other (Fig. 7.2a)
- move towards each other (Fig. 7.2b)
- move away from each other (Fig. 7.2c).

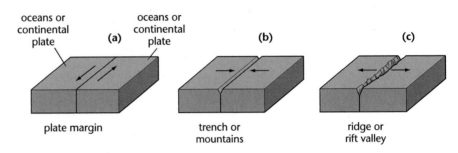

Fig. 7.2 Movement of plates.

(a) Plates are sliding past each other on the Californian coast in the USA (Fig. 7.2a). You may have heard of the San Andreas fault. Here plates are sliding past one another, causing earthquakes but *not* volcanoes.

(b) When an oceanic plate meets a continental plate (Fig. 7.2b), the denser oceanic plate is driven under the less dense continental plate. This is known as **subduction**.

The South Asian tsunami on 26 December 2004 was caused by subduction that resulted from a violent earthquake.

This causes:
- an oceanic trench
- a great deal of frictional heat, which melts the surrounding rock, resulting in volcanic activity.

When a continental plate meets another continental plate, a mountain is formed. This is how the Himalayas were created.

(c) When an oceanic plate moves away from another oceanic plate, the ocean widens (Fig. 7.2c). The gap is filled by rising magna, which rapidly cools and forms a ridge. The Mid-Atlantic ridge was formed in this way. Earthquakes and volcanoes occur along these ridges.

When a continental plate moves away from a continental plate, a rift valley is formed. The African rift valley was formed in this way. Earthquakes and volcanoes occur along these rifts.

At present there is no accurate way of predicting when earthquakes and volcanoes will occur.

7.2 Rocks as building materials

Materials used in building

Some rocks used for building materials are shown in the table below. These materials have been known for thousands of years.

Building material	Composition	Properties	Type of rock
limestone	calcium carbonate	easily cut; readily available; attacked by acid rain	Sedimentary rock formed from living organisms
sandstone	mainly silicon dioxide	comes in a variety of hardness and colours; absorbs water	Sedimentary rock formed from layers of sand
slate	mainly silicon dioxide	comes in a variety of colours and hardness	Metamorphic rock formed from clay mud

> You will have studied types of rocks and the rock cycle at Key Stage 3.

Because natural rocks are expensive to obtain, other materials such as mortar, cement, concrete and glass have been developed to replace them.

The table shows how some old materials have been replaced over the years.

Object	Old material	Replaced by	Advantage	Disadvantage
drainpipes	cast iron	PVC	does not break easily or corrode; no need to paint	can lose its shape over a period of time
rope	jute	nylon	stronger, more elastic	non-biodegradable
shopping bags	paper	polythene	has 'wet' strength	non-biodegradable
washing-up bowls and buckets	enamelled steel	polypropene	lighter, does not get dented	scratches
windows	glass	Perspex	less brittle, easily moulded	scratches

How materials can be modified

Plasticisers are substances added to materials to soften them.

The strength of concrete mixtures depends upon the ratio of water to cement. Concrete becomes weaker as more water is added to the cement. However, by adding less water, the concrete becomes difficult to mix and therefore unworkable. To overcome this problem, superplasticisers are added. Less water is needed to make the concrete, which makes it more workable and produces a high-strength concrete.

Plasticisers are also used to improve the properties of polymers. They are inserted between the chains of polymers to keep the chains apart. The chains can now easily slide over one another. This makes the polymer more flexible and softer, for example they make PVC softer and long-lasting.

Note that **plasticisers** in plastics soften the final product. Plasticisers in **concrete** soften the cement/water mixture before it sets.

The properties of polymers can also be changed by the following:

- **Increasing the chain length** – as the length of a polymer chain increases, it increases its physical properties such as melting point, density, ductility, tensile strength and hardness.

- **Cross-linking** – cross-links are short side chains of atoms linking two polymers together using covalent bonds. Cross-linking keeps the polymer chains apart. Vulcanisation is an example of cross-linking. Poly(isoprene) reacts with sulphur to make synthetic rubber. The cross-links join all the polymer molecules together.

  When rubber gets hot, the chains cannot move past each other. The cross-links do not break easily. This is why rubber tyres do not melt or get brittle when it is cold.

  The process of cross-linking is irreversible – the polymers are non-biodegradable. The only way of getting rid of them is to burn them.

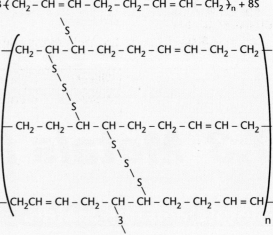

Fig. 7.3 Vulcanised rubber.

- **Increasing the crystallinity**. The term crystallinity is used to describe how the particles are arranged in a solid. If they are very orderly, then they have a high crystallinity. The linear chains in poly(ethene) can form ordered structures by lining up next to each other. This gives a very ordered structure that increases the density of poly(ethene) and also increases the melting point and the tensile strength. This form of poly(ethene) is referred to as HDPE (High Density Poly(Ethene)). If the poly(ethene)) chains are arranged randomly, then it is LDPE (Low Density Poly(Ethene). By looking at the structure of alloys, you may be able to work out why alloys have a low crystallinity.

In ancient times the Romans used blood as a plasticiser in their concrete.

The 'new' smell in cars and from synthetic carpets is due to the evaporation of plasticisers.

If you have had your hair permed, the protein polymers in your hair will have been cross-linked using sulphur bonds.

Perfect gemstones have a very high crystallinity

Elements in the Earth's crust

OCR B   C2c
AQA   C1.12.2
EDEXCEL 360   C1a6

The pie chart shows the **abundance of elements** by mass in the Earth's crust.

> **See if you can remember the symbols for all the elements in the pie chart.**

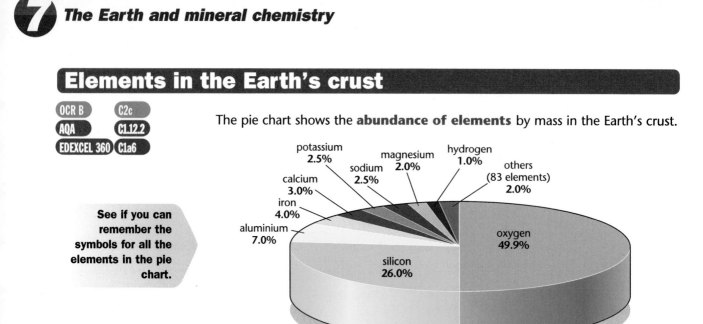

Fig. 7.4 The elements in the Earth's crust.

Most of the silicon and oxygen are present in the Earth's crust as the compound **silicon dioxide** (SiO_2). Sand is impure silicon dioxide. The presence of **transition element compounds** gives sand its different colours in different parts of the world.

Silicon dioxide has a giant structure of atoms held together by strong **covalent bonding**. The structure accounts for its properties:
● high melting point (over 1000°C) and a high boiling point (over 2000°C)
● very hard
● insoluble in water
● very poor conductor of electricity.

Silicon dioxide is found as quartz in granite, and is the main constituent of sandstone.

Glass is made by mixing sand, limestone and sodium carbonate together and melting the mixture. Coloured glass is made by adding transition metal compounds to the mixture.

7.3 *Metals from rocks*

Metals and non-metals

OCR A   C1.2
OCR B   C2d
AQA   C1.12.2
EDEXCEL 360   C1a6

Elements
Elements can be divided into **metals** and **non-metals**.
The table shows some elements, their symbols and their physical states at room temperature and pressure.

Metallic element	Symbol	State	Non-metallic element	Symbol	State
aluminium	Al	solid	argon	Ar	gas
calcium	Ca	solid	bromine	Br	liquid
copper	Cu	solid	carbon	C	solid
lead	Pb	solid	chlorine	Cl	gas
magnesium	Mg	solid	iodine	I	solid
sodium	Na	solid	neon	Ne	gas
tin	Sn	solid	nitrogen	N	gas
zinc	Zn	solid	oxygen	O	gas

The table shows the main differences in physical properties between metals and non-metals.

Metals, e.g. copper	Non-metals, e.g. nitrogen
shiny appearance (lustrous)	dull appearance if solid (non-lustrous)
solids (except for mercury)	either gases, volatile liquids or solids with low melting points (except for carbon)
malleable (can be hammered into different shapes without breaking); sonorous (make a ringing sound when struck); ductile (can be drawn into wires)	brittle if solid (easily broken when hammered)
high melting points and boiling points (there are some exceptions such as sodium, potassium and mercury)	low melting points and boiling points (except carbon)
good conductors of heat	poor conductors of heat
good conductors of electricity	poor conductors of electricity (except graphite)

Manufacture of metals

Most metals are found in the earth as deposits of **ore**.

 KEY POINT An ore is a rock that contains enough of the metal for it to be economically extracted.

The table lists some common ores.

Metal	Name of ore	Compound from which metal is extracted
aluminium	bauxite	aluminium oxide (Al_2O_3)
copper	malachite	basic copper(II) carbonate ($CuCO_3.Cu(OH)_2$)
iron	haematite	iron(III) oxide (Fe_2O_3)
mercury	cinnabar	mercury sulphide (HgS)
zinc	zinc blende	zinc sulphide (ZnS)
sodium	rock salt	sodium chloride (NaCl)

Some metals, such as silver and gold, are very unreactive. They occur as the metals themselves.

Extracting metals from ores

The method used to extract metals from their ores depends on the position of metals in the reactivity series.

You should be able to predict the method used to extract a metal from its ore given its position in the reactivity series.

Metals	Method of extraction
potassium (most reactive) sodium calcium magnesium aluminium	by electrolysis
zinc iron lead (least reactive)	by reduction with carbon

Extraction of aluminium

OCR B   C2e
AQA   C1.12.2
EDEXCEL 360   C1a6

We have seen that aluminium is extracted by electrolysis. Purified bauxite (aluminium oxide (Al_2O_3)) is dissolved in molten sodium aluminium fluoride (Na_3AlF_6) to form a mixture which has a relatively low melting point. The method used is shown in Fig. 7.5.

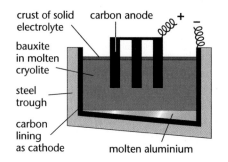

crust of solid electrolyte   carbon anode   +   –
bauxite in molten cryolite
steel trough
carbon lining as cathode   molten aluminium

Fig. 7.5 Apparatus for extracting aluminium.

The **electrodes** are made of **carbon**.

The **reactions** taking place are:

● at the **cathode**:

aluminium ions + electrons → aluminium

$$Al^{3+}(l) + 3e^- \rightarrow Al(l)$$

● at the **anode**:

oxide ions → oxygen + electrons

$$2O^{2-}(l) \rightarrow O_2(g) + 4e^-$$

The aluminium sinks to the bottom of the cell, where it is siphoned off.

The anode slowly burns away as the carbon reacts with oxygen to form carbon dioxide.

The cost of manufacture of aluminium from its ore is high. Recycled aluminium uses only about 5% of the energy required to extract it from bauxite.

Use of aluminium	Reasons for use
aircraft, ships, trains, cars	low density, strong, does not corrode
drinks cans	low density, strong, does not corrode
overhead power cables (with a steel core to strengthen them)	low density, does not corrode, good conductor of electricity
saucepans	low density, resists corrosion, good appearance, good conductor of heat

Don't forget to recycle your drinks cans.

Iron

OCR B C2c
AQA C1.12.2
EDEXCEL 360 C1a6

Iron is less reactive than carbon. It can be extracted from the oxides by reduction with carbon. **Haematite** (Fe_2O_3) is reduced in the blast furnace to make iron.

$$iron(II)\ oxide + carbon \rightarrow iron + carbon\ monoxide$$
$$Fe_2O_3(s) + 3C(s) \rightarrow 2Fe(s) + 3CO(g)$$

Iron from the blast furnace contains about 96% iron. The impurities in the iron make it brittle and therefore it has limited uses.

If all the impurities were removed from iron, the pure iron would be too soft for many uses.

Pure iron has a regular arrangement of atoms, with layers that can slide over each other.

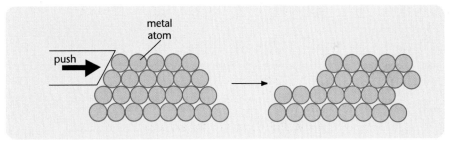

metal atom

push

Fig. 7.6 Iron has layers that slide over each other.

Most iron is converted into steels. Steels are **alloys** since they are mixtures of iron with carbon and other metals. The different sized atoms distort the layers in the structure of the pure metal, making it more difficult for them to slide over each other. This is why alloys are harder than pure metals.

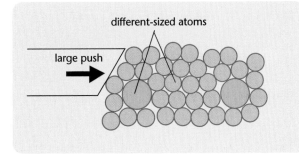

different-sized atoms

large push

Fig. 7.7 Steel is an alloy, and the different sized atoms prevent the layers moving easily.

Rusting

Rusting is the **corrosion** of iron or steel to form hydrated iron(III) oxide. It is an oxidation reaction and an exothermic reaction.

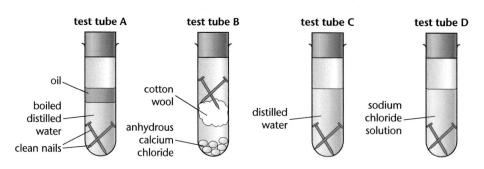

Fig. 7.8 Experiment to show rusting.

The experiment shown in Fig. 7.8 can be used to show that both air and water are necessary for rusting.

Anhydrous calcium chloride removes water from the test tube. Boiling distilled water removes air. (Remember that the active part of air is oxygen.) Oil stops any air from dissolving back in the boiled water.

The table shows the condition inside the test tubes.

Test tube	Conditions		Observation
	Is water present?	**Is air present?**	
A	yes	no	no rusting
B	no	yes	no rusting
C	yes	yes	nails rusted
D	yes	yes	nails very badly rusted

After one week, rusting will have occurred in test tubes **C** and **D** but not in test tubes **A** and **B**. The most heavily rusted nails are in test tube **D**.

Therefore air and water together are necessary for rusting. The presence of sodium chloride speeds up the rusting process. Acid rain also speeds up rusting.

Car bodies are made of steel and therefore they rust. They will rust more quickly if you live by the sea.

This table compares iron and aluminium.

		iron	aluminium
differences	**magnetism**	magnetic	non-magnetic
	rusting	rusts	does not rust
	density	more dense than aluminium	less dense than iron
similarities	**malleability**	malleable	malleable
	electricity and heat	good conductor	good conductor

Salt water and acid rain accelerate the rusting process. Aluminium does not corrode in moist conditions because it has a protective layer of aluminium oxide which does not flake off the surface.

Although car bodies made of aluminium would not rust and would be much lighter than a car made of steel, they would be much too expensive to manufacture.

Copper

OCR B C2d
AQA C1.12.2
EDEXCEL C1a6

Copper has properties that make it useful for electrical wiring and plumbing.

Pure copper is a good **conductor**. Copper is purified by **electrolysis** using the cell shown in Fig. 7.9.

Pure copper is used as the cathode, impure copper as the anode and copper(II) sulphate as the electrolyte.

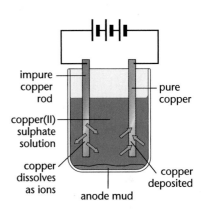

Fig. 7.9 How copper is purified

> **KEY POINT**
>
> During electrolysis copper from the anode goes into solution as copper ions. Copper ions from the copper(II) sulphate are deposited on the cathode.
>
> The changes taking place are:
>
> At the anode $\qquad$ copper $\rightarrow$ copper ions + electrons
> $$Cu(s) \rightarrow Cu^{2+}(aq) + 2e^-$$
>
> At the cathode $\quad$ copper ions + electrons $\rightarrow$ copper
> $$Cu^{2+}(aq) + 2e^- \rightarrow Cu(s)$$
>
> Impurities in the copper collect under the copper anode.

The impurities contain valuable metals such as silver and gold.

The supply of copper-rich ores is limited. We need new ways of extracting copper from low-grade ores to limit the environmental impact of traditional mining. Most of the copper we buy nowadays has been recycled from old copper wires and pipes.

7.4 Alloys

Alloys and their properties

OCR B C2d
AQA C1.12.2

> **KEY POINT**
>
> An alloy is a mixture of metals with other elements (usually metals but sometimes non-metals such as carbon).

Alloys can be designed to have properties for specific uses. **Low-carbon steels** (about 0.1% carbon) are easily shaped; **high-carbon steels** (about 1% carbon) are hard; and stainless steels are resistant to corrosion.

Steels are harder and stronger than iron and less likely to corrode.

Many metals in everyday use are alloys. Pure copper, gold and aluminium are too soft for many uses and so are mixed with small amounts of similar metals to make them harder for everyday use. **Smart alloys** can return to their original shape after being deformed.

The table shows some common alloys and their composition.

Alloy	Composition (percentage of elements by mass)	Use	Reason
amalgam	mercury (50%), silver (35%), plus varying amounts of other elements such as copper, tin and zinc	dentistry	easily moulded
brass	copper (66–70%), zinc (30–34%)	statues, jewellery	resistant to corrosion
bronze	copper (90%), tin (10%)	machinery parts	hard, easily cast, resistant to corrosion
duralumin	aluminium (95%), copper (4%), magnesium (1%)	aircraft	strong, low density
nitinol	nickel (56%), titanium (44%)	aid to mending bones	memory alloy
solder	lead (67%), tin (33%)	joining metal surfaces	low melting point
stainless steel	iron (73%), chromium (18%), nickel (8%), carbon (1%)	cutlery	does not corrode or stain
steel	varies iron (≈99%), carbon (≈1%)	cars and lorries	malleable, relatively cheap

Alloys have **chemical properties** similar to those of the elements they contain, but they have different **physical properties**. For example, lead melts at 327°C and tin at 232°C, but solder melts at 183°C.

Pure gold (called 24 carat gold) is very soft. It is alloyed with silver and copper to make it harder and stronger.

 KEY POINT Carat means a 24th; thus in 9 carat gold, 9 parts in 24 parts are gold. The rest is silver and copper.

Transition metals are good conductors of heat and electricity and can be bent or hammered into shape. They are useful as structural materials and for making things that must allow heat or electricity to pass through them easily.

Recycling metals

It is predicted that many **metal ores** will run out within the next hundred years. The metals most at danger are **tin** and **lead**.

Reserves of raw metals may last longer:
- if **new ore deposits** are discovered. Satellites are used to help find new deposits. If a suitable ore is found, it has to be dug out of the ground,

crushed and then the metal extracted using a method that depends on the ore's reactivity. The company has to get rid of the waste. All these processes are very expensive.

- if **substitutes** are found to replace metals. It would then be possible to use existing metals more sparingly. For example, metal parts in cars could be replaced with plastic or ceramic parts. In communications, optical fibres are replacing the copper wires used in communication instruments.

- if metals are **recycled**. This occurs already – scrapped cars are crushed and melted for reuse; aluminium cans are collected and the lead electrodes from lead batteries are collected. You can help by recycling all the objects you use that contain metals, including the batteries you use in your various instruments.

7.5 Salts

Preparing soluble salts

EDEXCEL 360 C1a6

There are five methods of preparing soluble salts:

- acid + metal

hydrochloric acid + magnesium → magnesium chloride + hydrogen
$$2HCl(aq) + Mg(s) \rightarrow MgCl_2(aq) + H_2$$

- acid + metal oxide

sulphuric acid + copper(II) oxide → copper(II) sulphate + water
$$CuO(s) + H_2SO_4(aq) \rightarrow CuSO_4(aq) + H_2O(l)$$

- acid + metal hydroxide

nitric acid + lead(II) hydroxide → lead(II) nitrate + water
$$2HNO_3(aq) + Pb(OH)_2(s) \rightarrow Pb(NO_3)_2(aq) + H_2O(l)$$

- acid + metal carbonate

hydrochloric acid + barium carbonate → barium chloride + water + carbon dioxide
$$2HCl(aq) + BaCO_3(s) \rightarrow BaCl_2(aq) + H_2O(l) + CO_2(g)$$

The method of preparation is the same in each case. The method is summarised in Fig. 7.10.

- acid + alkali

hydrochloric acid + sodium hydroxide → sodium chloride + water
$$HCl(aq) + NaOH(aq) \rightarrow NaCl(aq) + H_2O(l)$$

> **An alkali is a metal oxide or metal hydroxide that is soluble in water.**

This reaction is known as **neutralisation**.

> **KEY POINT** **A neutralisation reaction is a reaction between an acid and an alkali, in the correct proportions to produce a neutral salt.**

The method used depends upon factors such as speed of reaction, availability of chemicals and cost.

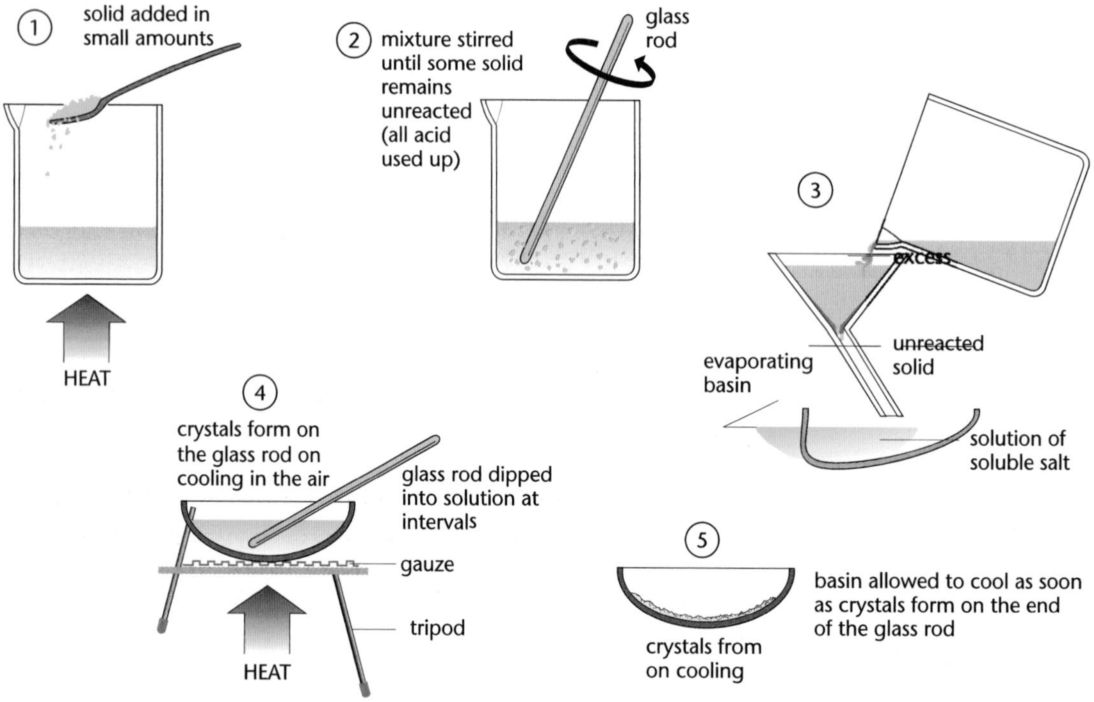

1. solid added in small amounts

HEAT

2. mixture stirred until some solid remains unreacted (all acid used up)

glass rod

3.

evaporating basin

excess

unreacted solid

solution of soluble salt

4. crystals form on the glass rod on cooling in the air

glass rod dipped into solution at intervals

gauze

tripod

HEAT

5.

crystals from on cooling

basin allowed to cool as soon as crystals form on the end of the glass rod

Fig. 7.10 How to prepare a salt.

A **titration method** is used if the hydroxide is an alkali, such as potassium hydroxide (KOH), sodium hydroxide (NaOH) or ammonia solution (NH_3(aq)).

The nitric acid is placed in the beaker using a pipette. Potassium hydroxide solution is added from the burette. An indicator such as litmus, phenolphthalein or methyl orange could be used to detect the end-point.

The solution is evaporated to crystallisation point, cooled and filtered. The crystals are washed and then dried between pieces of filter paper.

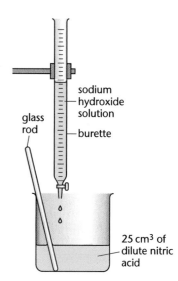

glass rod

sodium hydroxide solution

burette

25 cm³ of dilute nitric acid

Fig. 7.11 Preparing a salt using the titration method.

Potassium nitrate is used as a fertiliser. This method can also be used for making the fertiliser ammonium nitrate:

nitric acid + ammonia solution → ammonium nitrate

$HNO_3(aq) + NH_3(aq) → NH_4NO_3(aq)$

The solution is evaporated to crystallisation point, cooled and filtered. The crystals are washed and then dried between pieces of filter paper.

Preparation of an insoluble salt

 KEY POINT   **Insoluble salts are made by the process of precipitation.**

Two solutions each containing half of the salt to be prepared are mixed together. The salt is then precipitated.
Lead(II) carbonate can be prepared by mixing together lead(II) nitrate solution and sodium carbonate solution.

Common insoluble salts include silver chloride, barium sulphate and most carbonates (except carbonates of sodium, potassium and ammonium).

lead(II) nitrate + sodium carbonate → lead(II) carbonate + sodium nitrate

$Pb(NO_3)_2(aq) + Na_2CO_3(aq) → PbCO_3(s) + 2NaNO_3(aq)$

In order to get a pure sample of lead(II) carbonate the mixture is filtered, washed with distilled water and dried.

KEY POINT   **This type of reaction is sometimes called double decomposition and is represented by the equation AX + BY → AY + BX**

HOW SCIENCE WORKS

Smart alloys

Smart alloys have the ability to return to a pre-set shape. They can be twisted and bent but they still return to their original shape.

Nitinol is a smart alloy. It is named after the elements it contains and where it was discovered: Ni (nickel); Ti (titanium) and NOL (Naval Ordnance Laboratory). Its composition is 55–56% nickel and 44–45% titanium. However, small changes to these amounts cause changes in the properties of the nitinol.

There are two types of nitinol:

- 'SuperElastic' which is resistant to kinking and is very flexible. You may know someone who owns a pair of spectacles with nitinol frames. The frames can be twisted and bent but they still return to their original shape.
- 'Shape Memory', which returns to a pre-set shape when heated. Even a small temperature change can bring about this change. Imagine owning a car that returned to its original shape after an accident. You might be able to just take it through a car wash to return the car to its original shape.

Nitinol properties are explained by the fact that it can exist in two different forms – martensite and austenite. Their different structures are represented below.

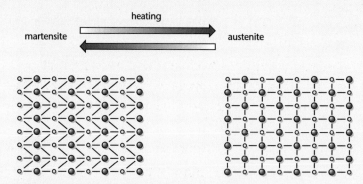

The yellow dots are nickel and the red dots titanium.

When nitinol is heated or cooled it changes from one form to the other. The pre-set form is martensite.

The European Space Agency's Rosetta spacecraft is due to land on comet Wirtanen some time in 2012. On board is a small strip of nitinol.

The space engineers needed a mechanism to open a seal on board the craft. Nitinol proved to be the answer. A nitinol strip has been bent out of shape and positioned near the seal. At the required moment the strip will be heated, so it will return to its original shape and break the seal.

Smart Memory Alloys (SMAs) are being used more and more. They can be used anywhere where there is a change in temperature. They are used as switches; various medical applications such as plates for broken bones, and for directing the airflow in heating units.

Exam practice questions

1. Which diagram represents an alloy **[1]**

(a)

(b)

(c)

(d)

2. What disaster happens when continental plates slide past one another? **[1]**
(a) earthquake and volcano **(c)** tsunami
(b) earthquake only **(d)** volcano only

3. Statues and monuments are often made of bronze. What are the two main metals
found in bronze? **[1]**
(a) copper and iron **(c)** iron and tin
(b) copper and tin **(d)** iron and zinc

4. 4. Complete the passage using only words from the following list:

ductile   electrical   harder   lower   malleable
metal   separate   soft   solder   useful

Dissolving an element, usually a metal, in a molten _____ forms an alloy. When the
solution solidifies on cooling, the solid produced is an alloy of the two metals. Alloys are
used where they have more _____ properties than the _____ components. One
of the properties of _____, which makes it more useful than the separate metals lead
and tin, is that it has a _____ melting point than either of the pure metals. This means
that it is quicker and easier to join pieces of other metals, such as copper wires in
_____ circuits.

Pure metals such as aluminium, iron and copper, are too _____ to be used for
construction purposes, but the addition of small amounts of other elements increases their
hardness. In a pure metal, the layers of atoms can move over each other relatively easily
and so the metal is soft, _____ (can be drawn out into a wire) and _____ (can
be beaten into different shapes). When atoms of another metal replace a small number of
atoms of the parent metal, they have the effect of making it more difficult for the layers of
atoms to move over each other. The alloy formed is _____.

[10]

Exam practice questions

5. Which acid would you use to make each of the following fertilisers?
   (a) ammonium sulphate   **(b)** ammonium phosphate   **[2]**

   Which alkali would you use to make each of the following fertilisers?
   (c) potassium nitrate   **(d)** ammonium sulphate   **[2]**

6. The diagram represents tectonic plates of the Earth's crust moving in the direction of the arrows.

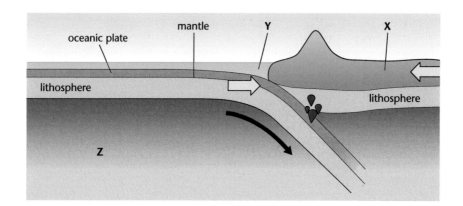

   (a) Name
        (i) the plate X   **(ii)** what is formed at Y   **(iii)** the region Z   **[3]**

   (b) What element is the most abundant in
        (i) the lithosphere   **(ii)** Z?   **[2]**

   (c) **(i)** Why does the oceanic plate go below plate X?
        (ii) What is this process known as?
        (iii) What natural disasters can happen at the plate margins?   **[4]**

7. A recycling company is deciding which metals to recycle.

Metal	Abundance of metal/metal ore in the Earth	Ease of extracting metal from Earth	Cost of preparing used metal for recycling
A	abundant	easy	moderate
B	rare	easy	low
C	moderate	difficult	high
D	rare	difficult	high

   State, with reasons:
   (a) the metal the company is *least* likely to recycle
   (b) the metal that could be gold.

Exam practice questions

8. The table below gives information about three metals: aluminium, gold and iron.

	Aluminium	Gold	Iron
symbol	Al	Au	
year discovered	1825	prehistoric	prehistoric
reaction with hydrochloric acid	slow, then very fast	no reaction	moderately fast
cost per 100 g (2003)	£0.97	£2380	£0.14
abundance %	7	0.000 000 4	4
% recycled	33	95	26

(a) What is the symbol for iron? **[1]**

(b) What is the reactivity of gold compared with aluminium and iron? **[1]**

(c) **(i)** Give one test that would show that aluminium, gold and iron were metals.

    (ii) Name the gas given off when iron reacts with hydrochloric acid. **[2]**

(d) **(i)** Suggest why aluminium is more expensive than iron although it is more abundant.

    (ii) Why is gold so expensive? **[2]**

(e) Suggest why gold and iron were known in prehistoric times. **[1]**

(f) In 1825, aluminium was made by reacting aluminium chloride ($AlCl_3$) with potassium. Write the word equation and symbol equation for this reaction. **[1]**

(g) **(i)** What is meant by the statement that 'metals are a finite resource'?

    (ii) Explain what is meant by the recycling of metals.

    (iii) Give *one* economical advantage and *one* environmental advantage for recycling metals.

    (iv) Give *one* reason for *not* recycling. **[5]**

9. Over the years plastics have been increasingly used as alternatives to other materials. Use the information in the table below to answer the questions.

Object	Old material	New material
chemical stoppers	cork	polythene
drain pipes	iron	PVC
lunch boxes	aluminium	polystyrene
water pipes	copper	polypropene
windows	glass	polycarbonates

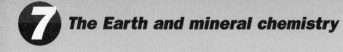

Exam practice questions

(a) Give a use of copper that will not be replaced by plastics. **[1]**

(b) **(i)** Give **one** advantage that polycarbonates have over glass.
(ii) Give **one** advantage that glass has over polycarbonates. **[2]**

(c) **(i)** Give **two** advantages that PVC has over iron.
(ii) What can be added to PVC to make it more flexible and softer? **[3]**

(d) Give **one** advantage that polythene stoppers have over cork stoppers. **[1]**

(e) Suggest why
(i) aluminium is suitable as cooking foil
(ii) polythene is unsuitable as cooking foil. **[2]**

(f) When PVC (polyl(vinylchloride)) burns, two of the products are carbon dioxide and water. Suggest two gases that might also be formed. **[2]**

(g) Polystyrene is a clear glasslike material. Expanded polystyrene boxes are used to protect breakable items. Suggest the difference between the structure of polystyrene and the structure of expanded polystyrene. **[1]**

The following topics are covered in this chapter:

- **Types of chemical reactions**
- **Identification of gases**
- **Rates of reaction**
- **Energy transfer in reactions**

8.1 Types of chemical reactions

Chemical reactions

> OCR A
> OCR B
> AQA
> EDEXCEL 360

} Most of these reactions occur in all the syllabuses.

There are many different types of chemical reactions. Some reactions fit into more than one heading.

Type of reaction	Definition	Example
Oxidation	When an element or compound adds oxygen or loses hydrogen. (It is the opposite of reduction.)	Magnesium is oxidised to magnesium oxide when it burns in air. $Mg(s) + O_2(g) \rightarrow 2MgO(s)$ Ammonia is reduced to nitrogen when it is heated with copper(II) oxide. $2NH_3(g) + 3CuO(g) \rightarrow N_2(g) + 3H_2O(l) + 3Cu(s)$
Reduction	When an element or compound loses oxygen or gains hydrogen. (It is the opposite of oxidation.) Usually oxidation and reduction occur together.	Copper(II) oxide is reduced to copper when it is heated in a stream of hydrogen gas. $CuO(s) + H_2(g) \rightarrow Cu(s) + H_2O(l)$ Oxygen is reduced to water when hydrogen burns in air. $2H_2(g) + O_2(g) \rightarrow 2H_2O(l)$
Thermal decomposition	The splitting up of a compound into simpler substances that do not recombine on cooling.	When copper(II) carbonate is heated, it thermally decomposes into copper(II) oxide and carbon dioxide. $CuCO_3(s) \rightarrow CuO(s) + CO_2(g)$
Cracking	The breaking down of an organic molecule into simpler substances.	Octane cracks into hexane and ethene when it is passed over a suitable catalyst. $C_8H_{16}(l) \rightarrow C_6H_{14}(l) + C_2H_4(g)$
Combustion (burning)	A reaction with oxygen which gives out heat energy.	When carbon burns in oxygen, it produces carbon dioxide and gives out energy. $C(s) + O_2(g) \rightarrow CO_2(g)$ This is also an example of oxidation.
Precipitation	A solid is formed when two soluble compounds are reacted.	Barium sulphate is precipitated when sodium sulphate solution is added to barium chloride solution. $BaCl_2(aq) + Na_2SO_4(aq) \rightarrow BaSO_4(s) + 2NaCl(aq)$
Displacement	A reaction when one element displaces another element.	Chlorine displaces bromine from a solution of potassium bromide. $Cl_2(g) + 2KBr(aq) \rightarrow Br_2(aq) + 2KCl(aq)$

8.2 Identification of gases

Tests for various gases

> Most of these tests for gases occur in all the syllabuses.

The table shows how gases can be identified by their properties.

Gas	Colour and odour	Test	Positive result
ammonia	colourless with a pungent smell	place a piece of moist red litmus paper at the mouth of the test tube moist *red* litmus paper	the red litmus paper turns blue
carbon dioxide	colourless and odourless	bubble gas through limewater carbon dioxide → limewater	a white precipitate is formed
chlorine	greenish-yellow with a pungent smell	place a piece of moist blue litmus paper at the mouth of the test tube moist *blue* litmus paper	the blue litmus paper turns red, and is then bleached
hydrogen	colourless and odourless	place a lighted splint at the mouth of the test tube pop — lighted splint — test tube holder	the lighted splint is extinguished with a 'pop' sound
oxygen	colourless and odourless	insert a glowing splint into the test tube flame — glowing splint	the glowing splint is rekindled (i.e. catches fire)

8.3 Rates of reaction

Reactions at different rates

 OCR B   C2g

> **KEY POINT** The rate of a reaction measures the rate of disappearance of reactant or the rate of appearance of a product.

> Some textbooks refer to 'rate of reaction' as 'speed of reaction'.

$$\text{Rate of reaction} = \frac{\text{Amount of reactant used up or Amount of product formed}}{\text{Time}}$$

Some reactions are very slow and other reactions are very fast.
- rusting is a slow reaction
- burning and explosions are very fast reactions.

Measuring rate of reaction

For practical reasons, reactions used in the laboratory for studying rate of reaction must not be too fast or too slow.

	Reaction in which amount of reactant is used up	**Reaction in which amount of product is formed**
Reaction studied	Marble chips and hydrochloric acid	Magnesium and hydrochloric acid
Equation and what is measured	$CaCO_3(s) + 2HCl(aq) \rightarrow CaCl_2(aq) + H_2O(l) + CO_2(g)$ Total loss in mass of calcium carbonate and HCl	$Mg(s) + 2HCl(aq) \rightarrow MgCl_2(aq) + H_2(g)$ Total volume of hydrogen given off
Apparatus	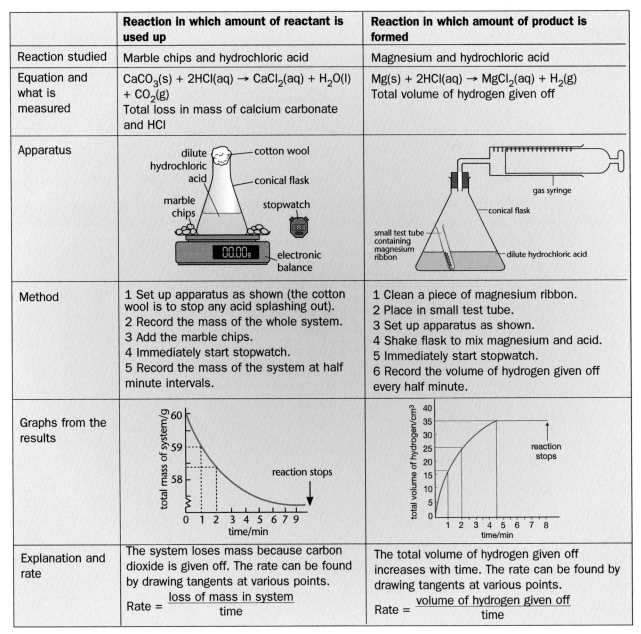	
Method	1 Set up apparatus as shown (the cotton wool is to stop any acid splashing out). 2 Record the mass of the whole system. 3 Add the marble chips. 4 Immediately start stopwatch. 5 Record the mass of the system at half minute intervals.	1 Clean a piece of magnesium ribbon. 2 Place in small test tube. 3 Set up apparatus as shown. 4 Shake flask to mix magnesium and acid. 5 Immediately start stopwatch. 6 Record the volume of hydrogen given off every half minute.
Graphs from the results		
Explanation and rate	The system loses mass because carbon dioxide is given off. The rate can be found by drawing tangents at various points. $$\text{Rate} = \frac{\text{loss of mass in system}}{\text{time}}$$	The total volume of hydrogen given off increases with time. The rate can be found by drawing tangents at various points. $$\text{Rate} = \frac{\text{volume of hydrogen given off}}{\text{time}}$$

A reaction stops when one of the reactants is used up.

Other methods of measuring rate of reaction include:
- colour changes
- formation of a precipitate
- temperature changes
- time taken for a given mass of solid to react.

Factors affecting rate of reaction

OCR B   C2g

The table compares some of the factors that affect the rate of a chemical reaction.

Factor	Types of reaction affected	Change made in the condition	Effect on rate of reaction
using a catalyst	slow reactions can be speeded up by adding a suitable catalyst	reduces amount of energy required for the reaction to take place	increases rate of reaction
concentration	all reactions	increases in concentration of one of the reactants	increases rate of reaction
light	wide variety of reactions including reactions with mixtures of gases, including chlorine and bromine	reaction in sunlight or UV light	greatly increases rate of reaction
particle size	reactions involving solids and liquids; solids and gases or mixtures of solids	makes solid particles smaller, e.g. use powdered form of the solid (s)	greatly increases rate of reaction
pressure	reactions involving gases	increases the pressure	greatly increases rate of reaction
temperature	all reactions	increases by 10°C decreases by 10°C	approximately doubles rate approximately halves rate

Explaining different rates using particle model

OCR B   C2g

The **collision theory** states that:
- chemical reactions can only occur when reacting particles collide with each other
- chemical reactions have a certain sufficient minimum energy called the **activation energy**.

The table shows how changing surface area, concentration and temperature affects the reaction rates.

Change	Before	After
increasing surface area ● more collisions ● reaction faster		
increasing concentration ● more collisions between particles ● more collisions leading to reactions ● reaction faster		
increasing temperature ● particles move faster ● more collisions and more energetic ● reaction faster		

Increasing pressure has the same effect.

Sunlight and UV light have the same effect.

Catalyst

KEY POINT

A catalyst is a substance that changes the rate of a chemical reaction but is not used up during the reaction.

● Different reactions need different catalysts.

● Catalysts are important in increasing the rates of chemical reactions used in industrial processes to reduce costs.

Rates of reaction in everyday life

We use rate of reaction in everyday life:

● the speed of cooking is increased by using a pressure cooker
● rusting is slowed down by covering iron objects with paint and oil
● tablets dissolve more quickly if they are powdered
● enzymes are biological catalysts; they speed up reactions that are essential to life.

Enzymes work best at 37°C (body temperature). At high temperatures, enzymes are destroyed (denatured).

Enzymes are also used:

● in the manufacture of alcohol from sugar
● in making cheese and yogurt
● in washing powders to break down protein stains such as blood.

8.4 Energy transfer in reactions

Endothermic and exothermic reactions

We all need energy to live. We get this energy from eating food. Some of the food is cooked, and cooking needs energy. If you have a gas cooker, then the energy will probably be supplied by burning natural gas (methane):

$$\text{methane} + \text{oxygen} \rightarrow \text{carbon dioxide} + \text{water} \quad \text{energy given off}$$
$$CH_4\,(g) + 2O_2 \rightarrow CO_2\,(g) + H_2O(l)$$

You will also have noticed that when gas burns, there are other forms of energy besides heat, namely light and sound.

> **KEY POINT**
>
> An **exothermic reaction** is a reaction in which energy is transferred into the surroundings (energy released).
>
> An **endothermic reaction** is a reaction in which energy is taken from the surroundings (energy absorbed).
>
> If there is a temperature rise, the reaction is exothermic.
>
> If there is a temperature fall, the reaction is endothermic.

Bond making and bond breaking

Bond making is an exothermic process and **bond breaking** is an endothermic process. During a chemical reaction some bonds are broken and new bonds are formed.

When natural gas is burned the following take place:

The four bonds in a methane molecule and the bond in each of the oxygen molecules is broken.

$$CH_4 \rightarrow C + 4H \text{ and}$$
$$2O_2 \rightarrow 4O \quad \text{energy required (endothermic)}$$

New bonds are then formed:
$$C + 2O \rightarrow CO_2 \text{ and}$$
$$4H + 2O \rightarrow 2H_2O \quad \text{energy given out (exothermic)}$$

The reaction is **exothermic**, so more energy was given out when the bonds were formed than the energy required to break the bonds.
In an exothermic reaction:
- the energy released when bonds are formed is greater than the energy absorbed when bonds are broken
- the energy released raises the temperature of the surroundings.

Examples of exothermic reactions are combustion, oxidation reactions and neutralisation reactions.

The reaction can be represented by an energy level diagram:

$CH_4(g) + 2O_2(g)$

energy given to the surroundings

$CO_2(g) + 2H_2O(l)$

> **Respiration is an exothermic process.**

Photosynthesis is an example of an **endothermic** reaction:

carbon dioxide + water → sugar + oxygen
$$6CO_2(g) + 6H_2O(l) \rightarrow C_6H_{12}O_6(s) + 6O_2(g)$$

The two bonds in the each of the carbon dioxide molecules are broken and the two bonds in each of the water molecules are broken:

$6CO_2 \rightarrow 6C + 12O$ and
$6H_2O \rightarrow 12H + 6O$ energy required (endothermic)

New bonds are formed:

> **Adding water to calcium oxide is called slaking. The reaction is very fast and gives out a great deal of heat energy. Calcium oxide is sometimes called quick lime.**

$6C + 12H + 6O \rightarrow C_6H_{12}O_6$ and
$12O \rightarrow 6O_2$ energy given out (exothermic)

The reaction is endothermic, so more energy was taken in when the bonds were broken than the energy given out forming new bonds.

In an endothermic reaction:
- the energy released when bonds are formed is less than the energy absorbed when bonds are broken.
- the extra energy is absorbed from the surroundings, so the temperature of the surroundings falls.

The reaction can be represented by an energy level diagram:

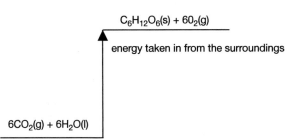

$C_6H_{12}O_6(s) + 6O_2(g)$

energy taken in from the surroundings

$6CO_2(g) + 6H_2O(l)$

Examples of endothermic reactions include thermal decompositions and dissolving salts in water.

If a reversible reaction is exothermic in one direction, it will be endothermic in the opposite direction. The amount of energy given out or taken in is the same.

The decomposition of calcium hydroxide to calcium oxide and water is an endothermic reaction:

$$Ca(OH)_2(s) \rightarrow CaO(s) + H_2O(g)$$

Adding water to calcium oxide to form calcium hydroxide is an exothermic reaction:

$$CaO(s) + H_2O(l) \rightarrow Ca(OH)_2(s)$$

Measuring the heat given out by a burning fuel

- Ethanol is placed in the spirit lamp and weighed.
- The temperature of the water in the copper calorimeter is measured.
- The spirit lamp is placed under the calorimeter and lit.
- The water is constantly stirred.
- When the temperature has risen by 10°C, the flame is put out.
- The highest temperature reached is measured.
- The spirit lamp is reweighed.

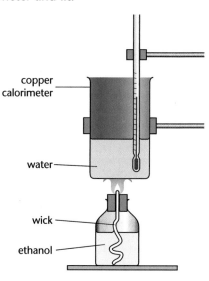

Fig. 8.1 How to measure the heat given out by a burning fuel.

copper calorimeter

water

wick

ethanol

4.2J is the energy required to raise 1g of water by 1°C.

The energy transferred to the water is calculated using the formula:

Energy transferred (in J) = mass of water heated (in g) × temperature (in °C) × 4.2

The result will be low because energy is lost heating the calorimeter and the surrounding air.

The energy output of ethanol in J/g is found using the formula:

energy/gram = $\dfrac{\text{energy transferred}}{\text{mass of fuel burned}}$

The experiment can be repeated using other fuels.

Structure and melting point

OCR B   C2.3

> **KEY POINT** Melting point is the temperature at which a solid changes to a liquid.

A pure substance at standard pressure (1 atmosphere) has a fixed **melting point**. If the substance is a mixture, it will melt over a range of temperatures. Melting point can be used to indicate the strength of forces between the particles in a substance. The lower the forces, the lower will be the melting point.

Helium has the lowest melting point (–272°C) and has the weakest forces between its atoms. Carbon in the form of graphite has the highest melting point (3675°C). Another form of carbon, diamond, has a melting point of 3550°C. Carbon must have very strong forces between its atoms.

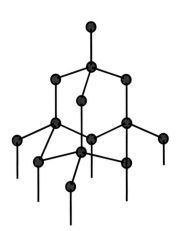

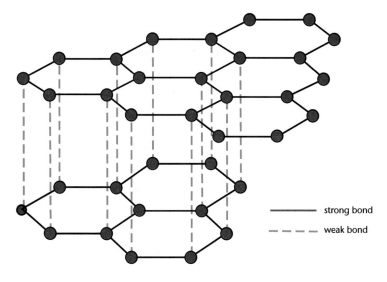

strong bond
- - - - weak bond

Fig. 8.2 The structure of diamond.　　**Fig. 8.3** The structure of graphite.

Usually metals and compounds of metals have higher melting points than non-metals and compounds of non-metals. Metal compounds have ionic bonds consisting of ions in a giant structure. The diagram illustrates the structure of sodium chloride.

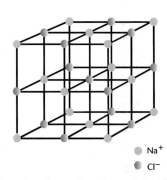

Na$^+$
Cl$^-$

Fig. 8.4 The lattice structure of sodium chloride.

Non-metal compounds have covalent bonds between their atoms to form molecules. The molecules are held together by weak van der Waals' forces.

Fig. 8.5 Iodine.

The two-dimensional structure of iodine is shown in Fig. 8.5. There are strong forces between each of the iodine atoms in a molecule of iodine, but weak van der Waals' forces between the molecules. The melting point of iodine is 114°C.

HOW SCIENCE WORKS

Life cycle assessment

Life cycle assessment (LCA) is the assessment of the environmental impact of a given product throughout its lifespan. Scientists use LCA to work out the most efficient method of manufacturing a substance and try to reduce any detrimental effect on the environment.

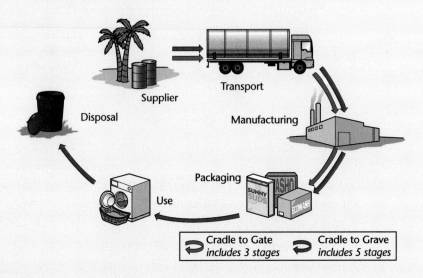

Some of the considerations made when building a factory are:

- the cost of energy
- sustainability
- impact on the environment when the product is made, used and disposed of.

Some of the environmental damages considered are global warming (greenhouse gases), acidification, ozone layer depletion, eutrophication and escape of poisonous pollutants.

LCA – aluminium

Aluminium is manufactured by the electrolysis of purified bauxite (aluminium oxide). Electrical energy is expensive. In order to reduce costs:

- aluminium smelters are built near hydroelectric power stations
- sodium aluminium fluoride is added to lower the melting point.

Sodium aluminium fluoride occurred as the ore cryolite. Cryolite has the dubious distinction of being one of the first ores to be used up. Nowadays the sodium aluminium fluoride has to be made. This adds to the cost of making aluminium.

Aluminium smelters emit enormous quantities of carbon dioxide (a greenhouse gas). They also emit carbon monoxide, sulphur dioxide and fluorides, the most dangerous pollutants in terms of public health and land damage. These gases kill the fauna and flora around the factory.

HOW SCIENCE WORKS

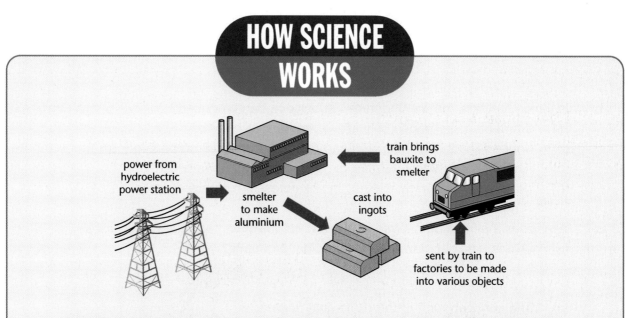

power from hydroelectric power station

smelter to make aluminium

cast into ingots

train brings bauxite to smelter

sent by train to factories to be made into various objects

Rich ores of bauxite are found in environmentally important areas of the world such as the Blue Mountains of Jamaica and the Amazon rain forest. There has been a great deal of controversy about the building of a gigantic hydropower station near Europe's largest glacier, Vatnajökull in Iceland, to supply electricity for an aluminium smelter. Despite strong arguments from environmentalists about the damage that will be caused, the manufacturing company got permission to build the plant.

The cost of obtaining aluminium from aluminium cans is 5% of the cost of manufacturing it from bauxite. All metal cans are recyclable, yet we recycle only about 33% of our cans; the rest is thrown into landfill sites. Aluminium foil and laminates (such as crisp packets) are more difficult to recycle. At present they are burned to release energy. During this process the aluminium is oxidised to aluminium oxide from which aluminium can be recovered.

The amount of waste that has to be removed in obtaining aluminium and iron can be seen in the table.

Metal	Ore	% of metals obtained	% waste
aluminium	bauxite	20	80
iron	haematite	30	70

The cost of removing the waste and where to put the waste are major environmental problems.

Exam practice questions

1. In which reaction has the underlined substance been reduced?
    (a) ammonia + <u>hydrogen chloride</u> → ammonium chloride
    (b) copper(II) sulphate + <u>sodium carbonate</u> → copper(II) carbonate + sodium sulphate
    (c) hydrogen sulphide + <u>sulphur dioxide</u> → sulphur + water
    (d) potassium hydroxide + <u>nitric acid</u> → potassium nitrate + water   **[1]**

2. Why does the reaction between marble chips (calcium carbonate) and dilute
    hydrochloric acid get slower as the reaction proceeds?   **[1]**

    $$CaCO_3(s) + 2HCl(aq) → CaCl_2(aq) + CO_2(g) + H_2O(l)$$

    (a) The concentration of the dilute acid decreases.
    (b) The mixture becomes saturated with calcium chloride.
    (c) The surface area of the marble chips becomes larger.
    (d) The temperature of the solution increases.

3. The diagram shows ammonium chloride being heated. What does the clear region
    contain?   **[1]**
    (a) ammonia and chlorine
    (b) ammonia and hydrogen
         chloride
    (c) hydrogen chloride and
         nitrogen
    (d) hydrogen, chlorine and
         nitrogen

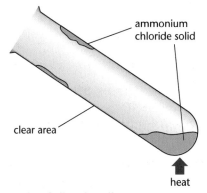

ammonium
chloride solid

clear area

heat

4. Complete the passage using only words from the following list

    faster   larger   loss   marble   salt   the same   smaller
    temperature   time   volume

    The rate of reaction between excess _____ chips (calcium carbonate) and dilute
    nitric acid can be found by measuring the _____ in mass of the apparatus, or
    the _____ of carbon dioxide given off and the _____ that each measurement
    is taken. The _____ formed in this reaction is calcium nitrate.

    The rate of reaction could be made faster by increasing the _____. It can also be
    increased by making the marble chips _____.

    If more marble chips had been used, the volume of carbon dioxide given off in the same
    time would have been _____. If concentrated nitric acid had been added to the same
    volume of dilute nitric acid before reacting with the same mass of marble chips,
    the reaction would have been _____ and the volume of carbon dioxide given off
    would have been _____.   **[10]**

Exam practice questions

5. Hydrogen reacts with fluorine to form hydrogen fluoride. The reaction is exothermic:

    $$H_2(g) + F_2(g) \rightarrow 2HF(g)$$

    During this reaction covalent bonds are broken and new covalent bonds are formed.
    (a) What type of energy change occurs when bonds are broken?   [1]
    (b) What type of energy change occurs when bonds are formed?   [1]
    (c) Which is the larger of the energy changes in the reaction above, bond making
         or bond breaking? Explain your answer.   [2]

6. The apparatus shown was used to
    determine the heat energy given out when
    1 g of different alcohols were burned in air.

    (a) Give two reasons why this apparatus
         gives more accurate results than the
         apparatus shown in Fig. 8.1.   [2]
    (b) List the readings that need to be taken
         to measure the energy change.   [5]
    (c) Why is a copper tube used in this
         experiment and not a polythene tube?   [2]
    (d) When each experiment is finished a
         close fitting cap is put on the burner. Why?   [1]
    (e) The burning of alcohols is an exothermic
         reaction. What can you deduce from this fact?   [2]
    (f) The table below lists the results taken for
         the combustion of five different alcohols.

Alcohol	Formula	Heat energy kJ/g
methanol	CH_3OH	23.0
ethanol	C_2H_5OH	30.0
propanol	C_3H_7OH	33.5
butanol		36.0
pentanol	$C_5H_{11}OH$	
hexanol	$C_6H_{13}OH$	40.0

    Suggest
         (i)   the formula of butanol
         (ii) the missing value for the heat energy of pentanol.   [2]
    (g) What further information is required before you can decide which is the best
         alcohol to use as a fuel in your home.   [1]

Exam practice questions

7. An experiment was carried out to investigate the rate of reaction between magnesium and sulphuric acid.

    0.07 g of magnesium ribbon was reacted with excess dilute sulphuric acid. The volume of gas produced was recorded every 10 seconds. The experiment was repeated using the same mass of magnesium and excess sulphuric acid. The results are shown below.

Experiment 1		Experiment 2	
Time in seconds	Total volume in cm^3	Time in seconds	Total volume in cm^3
0	9	0	0
10	43	10	34
20	66	20	44
30	76	30	66
40	79	40	68
50	81	50	68

   (a) Draw a diagram of the apparatus used to carry out the experiment. Use a gas syringe to measure the volume of gas given off. **[3]**

   (b) **(i)** Use these results to plot a graph of these results with the total volume of hydrogen on the *y*-axis.

   (ii) Describe how you used these results to plot the graph. **[6]**

   (c) When was the reaction fastest? Explain your answer. **[2]**

   (d) How long did it take for 0.07 g of magnesium to completely react? **[1]**

   (e) At what time was 0.02 g of magnesium left unreacted? **[1]**

   (f) On your graph sketch the curve you would expect to obtain if 0.07 g of powdered magnesium were used under the same conditions. **[2]**

The following topics are covered in this chapter:

- **Radioactive emissions**
- **Dangers of radiation**
- **Uses of radioactivy**
- **Changes in the nucleus**
- **Nuclear reactors**

9.1 Radioactive emissions

Alpha, beta and gamma radiation

OCR A — P3.1
OCR B — P2d
AQA — P1.13.6

A **radioactive nucleus** is **unstable** and will emit radiation. There are three main types:

- **Alpha (α)** is strongly ionising radiation, but it only travels a few centimetres in air and is stopped by a thin sheet of paper.
- **Beta (β)** is ionising radiation that penetrates card or several sheets of paper, but is stopped by a 3mm thick sheet of aluminium or other metal.
- **Gamma (γ)** is weakly ionising radiation that is very penetrating. It is reduced significantly by a thick lead sheet or blocks of concrete.

> **KEY POINT**
>
> There are three types of radioactive emissions: alpha, beta **and** gamma radiation. Alpha radiation is the most ionising, gamma radiation is the most penetrating.

Alpha and **beta** radiation is deflected by **magnetic fields** and **electric fields**, but **gamma** radiation is not affected.

Background radiation

OCR A — P3.2
OCR B — P2d
AQA — P1.13.6

Radioactive materials occur naturally, and can also be made artificially. **Cosmic rays** from space make some of the carbon dioxide in the atmosphere radioactive. The carbon dioxide is used by plants and enters food chains. This makes all living things radioactive. Some rocks are also radioactive.

We receive a low level of radiation from these sources all the time. It is called **background radiation.** Background radiation comes from:

- space (cosmic rays) and the Sun
- building materials, rocks (e.g. granite) and soil
- radioactive nuclei in all plants and animals
- medical and industrial uses of radioactive materials
- 'leaks' from radioactive waste and nuclear power stations.

Background radiation is radioactive emissions from nuclei in our surroundings. Do not confuse this with radiation from other sources, such as radiation from mobile phones.

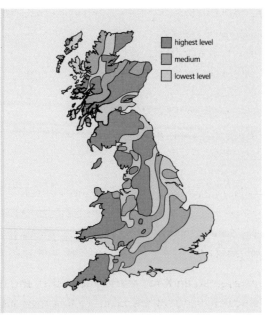

legend:
- highest level
- medium
- lowest level

Fig. 9.1 In some parts of the country the rocks are more radioactive than in others and there is a higher level of background radiation.

9.2 *Dangers of radiation*

Ionising radiation

OCR A — P3.2
OCR B — P2c
AQA — P1.13.6
EDEXCEL 360 — P1b11

Gamma rays, X-rays and ultra-violet radiation are the three types of electromagnetic radiation with high enough frequency to cause ionisation.

Ionising radiation is radiation that has enough energy to break molecules or atoms into charged particles called **ions**. The molecules or atoms lose **electrons** in a process called **ionisation**. The ions can then take part in chemical reactions. This is how ionising radiation damages **living cells**. It kills them, or damages the **DNA** in the cell so that the cell **mutates** (changes) into a **cancer** cell, which then grows in an uncontrolled manner.

Ionising radiation includes alpha, beta and gamma radiation from the radioactive nuclei and also X-rays and ultra-violet radiation.

Risk and safety

OCR A — P2.5/3.2/3.4
OCR B — P2d
AQA — P1.13.6
EDEXCEL 360 — P1b11

It is **not possible to predict** which cells will be damaged by exposure to radiation, and it is not possible to say who will get cancer. Scientists have studied the survivors of incidents where people have been exposed to ionising radiation. They measured the amount of exposure and recorded how many people later suffered from cancer.

They introduced **radiation dose**, measured in **sieverts**, which is a measure of the **possible harm done to the body**. Radiation dose depends on the **type of radiation**, the **time of exposure** and how **sensitive the tissue** exposed is to radiation. Scientists can then give a figure for the **risk** of cancer developing.

To reduce the risk to living cells, radioactive materials must be handled **safely**:

- wear protective clothing
- keep a long distance away (use tongs to handle sources)
- keep the exposure time short
- sources should be shielded and labelled with the radioactive symbol.

Fig. 9.2 Radioactive hazard symbol.

These precautions keep the dose as low as possible. Radioactive materials are used, for example, in hospitals and nuclear power stations. Employers must keep the dose for their employees **as low as reasonably achievable**: this is known as the **alara** principle.

You may have had an X-ray photograph taken and noticed that the staff move behind a screen or out of the room, so that they are not exposed to X-rays every time the X-ray machine is used. Employees at nuclear powers stations, and other places where nuclear sources are used, may wear a **film badge** to monitor the exposure to radiation. Airline employees working as flight crew are monitored too, because the exposure to cosmic rays is increased if you fly at high altitudes in the thin atmosphere of the Earth.

9.3 Uses of radioactivity

Uses of radioactive sources

Smoke detectors

Radioactive sources that emit **alpha radiation** are used in **smoke detectors**. The alpha radiation from the source crosses a small gap and is picked up by a detector. If smoke is present, the alpha radiation is stopped by the smoke particles. No radiation reaches the detector and an alarm is sounded.

Beta and gamma radiation are unsuitable because they pass through the smoke. For each use, the radiation is chosen depending on the range and the absorption (see p. 146).

Paper thickness detectors

Radioactive sources that emit **beta radiation** are used in **paper thickness detectors**. Figure 9.3 shows how the paper thickness can be monitored. Some of the beta radiation is **absorbed** by the paper sheet. If the sheet is too thick, less beta radiation is detected and the pressure of the rollers is increased. If the sheet is too thin more beta radiation is detected and the pressure is reduced.

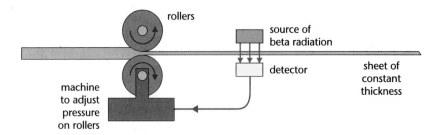

Fig. 9.3 Using a beta source to control paper thickness

Tracers

Radioactive sources that emit **beta radiation** are used as **tracers**. Because a tracer is radioactive, detectors can be used to track where it goes. If a tracer is added to sewage at an ocean outlet, or as it enters a river, then the movement can be traced. Leaks in power station heat exchangers can be tracked. The source used is carefully selected to be one with a radioactivity that will fall to zero quickly after the test is done.

Sources that emit **gamma radiation** are used in **medical tracers**. The patient drinks, inhales or is injected with the tracer which is chosen to target the organ doctors want to examine. For example, radioactive iodine is taken up by the thyroid gland, which can then be viewed using a **gamma camera** that detects the gamma radiation passing out of the body. Sources that emit a **higher dose of gamma radiation** are used in the same way for **treating cancer** by killing the cancer cells.

Sterilisation

Radioactive sources that emit **gamma radiation** are used to produce a beam of gamma rays that will:
- **sterilise equipment**, such as surgical instruments, by destroying microbes
- extend the shelf-life of perishable **food**, by destroying microbes.

The food and equipment do not become radioactive because they do not change the nuclei, so no particles of radioactive material are in the food. The rays – like light rays – are absorbed, or pass through, and have gone in less than a second.

Non-destructive testing

Another use of **gamma radiation** sources is **non-destructive testing**. An aircraft wing can be examined for minute cracks by placing a strong gamma source on one side and a detector on the other – in a similar way to using X-rays to check for a broken bone.

9.4 Changes in the nucleus

The nucleus and isotopes

OCR A P3.1
OCR B
AQA P1.13.6
EDEXCEL 360

The **atom** is mostly empty space with almost all the **mass** concentrated in the small **positively charged nucleus** at the centre. The **nucleus** of an atom contains two types of particle:

- **neutrons**, which have no charge
- **protons** – each proton has a single positive charge.

The **nucleus** is surrounded by orbiting **electrons** which have very little mass. Each **electron** has a negative charge. A neutral atom has an equal number of protons and electrons.

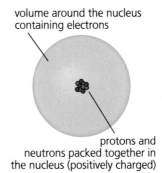

volume around the nucleus
containing electrons

protons and
neutrons packed together in
the nucleus (positively charged)

Fig. 9.4 Arrangement of neutrons, protons and electrons in an atom.

> **KEY POINT**
> The atomic number or proton number, **Z**, is the number of protons in the nucleus. The mass number or nucleon number, **A**, is the total number of protons and neutrons in the nucleus.

Isotopes

The number of protons in the nucleus determines the element. For example, a hydrogen nucleus always has one proton and a carbon nucleus always has six protons. The number of neutrons in the nucleus can vary. An **isotope** is a nucleus of an element with the **same number** of **protons** but different **number** of **neutrons**. Some isotopes are stable but others are unstable. Unstable isotopes are radioactive because they emit nuclear radiation in the process called **radioactive decay**.

> **KEY POINT**
> Isotopes of an element have the same number of protons in the nucleus, but different numbers of neutrons.

Isotopes of the same element have exactly the same chemical properties, but they have different density and nuclear stability.

The nucleus is given a symbol $^A_Z X$ where A is mass number, Z is the proton number and X is the chemical symbol for the element. For example, $^{12}_6 C$ is a stable isotope of carbon with six protons and six neutrons.

Alpha emission

Alpha emission is when **two protons and two neutrons** leave the nucleus as one particle. This alpha particle is identical to a helium nucleus so has the symbol: $_{2}^{4}\text{He}$. The new nucleus has a mass number that has decreased by four and an atomic number that has decreased by 2. For example, the isotope of radon gas, radon-220, decays by alpha emission. The decay equation is:

$$_{86}^{220}\text{Rn} \rightarrow \, _{84}^{216}\text{Po} + \, _{2}^{4}\text{He}$$

> A beta particle is an electron from the nucleus – not an orbital electron.

Beta emission

Beta emission occurs when a **neutron** decays to a **proton** and an **electron**. This electron leaves the nucleus as a beta particle. The atomic number increases by one and the mass number is unchanged. The beta particle is a high-energy electron and has the symbol: $_{-1}^{0}\text{e}$.

$_{6}^{14}\text{C}$ is a radioactive isotope of carbon with six protons and eight neutrons. This nucleus emits beta radiation and forms a nitrogen nucleus. The decay equation is:

$$_{6}^{14}\text{C} \rightarrow \, _{7}^{14}\text{N} + \, _{-1}^{0}\text{e}$$

> **KEY POINT**
>
> An alpha particle is two protons and two neutrons – a helium nucleus. A beta particle is a high-energy electron from the nucleus. Gamma radiation is a high-frequency and short wavelength electromagnetic wave.

Radioactive decay and half-life

A radioactive nucleus is unstable and emits nuclear radiation. This process is called **radioactive decay**. It is not possible to predict when an individual nucleus will decay, nor is it possible to make it happen by a chemical or physical process (for example by heating it). The **decay** is **random**.

A radioactive source contains millions of nuclei. The number of nuclei decaying per unit time is called the **activity** of the source. The activity depends on two things:

- the type of isotope – some isotopes are more stable than others.
- the number of undecayed nuclei in the sample – double the number of nuclei and, on average, there will be double the number of decays per second.

Over a period of time the activity of a source gradually dies away.

The **half-life** of an isotope is the average time taken for half of the active nuclei to decay. Technetium – 99m (Tc-99m) decays by gamma emission to Technetium-99 (Tc-99) with a half-life of six hours. After six hours, on average, only half of the Tc-99m nuclei remain active. After another six hours, on average, only one quarter of the nuclei are active. This is shown on the graph in Fig 9.5. This pattern is the same for all isotopes but the value of the half-life is different. Carbon-14 has a half-life of 5730 years, but some isotopes have a half-life of less than a second.

Tc-99m is one of the most widely used radioactive isotopes in medicine. It is used to diagnose problems in many organs. The half-life of six hours means that the radiation lasts only long enough for the isotope to travel to the organ being investigated. Its activity decreases rapidly and cannot be detected after a few days. Tc-99m does not occur naturally: it is made in reactors.

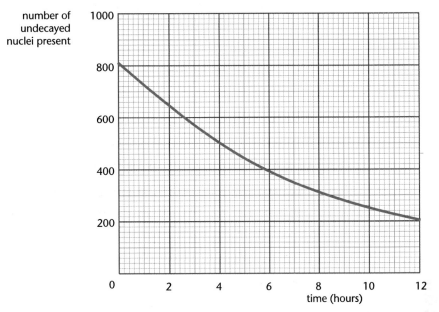

Fig. 9.5 The decay of a sample radioactive nuclei with a half-life of six hours.

9.5 Nuclear reactors

Nuclear fission

OCR A   P3.3
OCR B   P2c

If a nucleus of uranium-235 absorbs a neutron, it becomes very unstable and can split into two nuclei of about equal size, and two or three neutrons. This process is called **nuclear fission**.

When this happens a lot of nuclear energy is released, about a **million times more** than the energy released in a chemical reaction.

The neutrons released can strike more uranium nuclei and cause more fission reactions – which in turn produce more neutrons, and so on. This is called a **chain reaction**.

If the fission reaction runs out of control it is an atomic bomb, but if the process is controlled, the energy released can be used to generate power. This is how a **nuclear reactor** in a **power station** works.

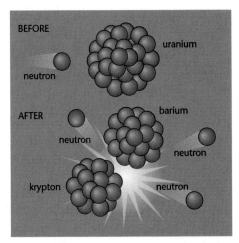

Fig. 9.6 A neutron absorbed by a uranium nucleus causes nuclear fission.

The fuel rods contain uranium-235 and are put in the reactor. The reaction can be controlled or stopped by using a material that absorbs neutrons. The neutron absorbing material is made into **control rods**. The control rods are moved into the reactor to absorb neutrons, and slow or stop the reaction, and out of the reactor to increase the reaction.

The energy heats up the fuel rods and control rods. A **coolant** is circulated to remove the heat from the reactor. When the coolant has been heated it is used to heat water to steam for the power station. When the coolant is cool it circulates through the reactor again.

Waste disposal

OCR A P3.3
OCR B P2d

Radioactive waste is dangerous to living things and must be carefully disposed of. The half-life of some isotopes is thousands, or millions, of years, so radioactive material must be disposed of in a way that will keep it safely contained for thousands of years.

There are three types of radioactive waste:
- **low-level waste** – for example, used protective clothing.
- **intermediate-level waste** – for example, material from reactors.
- **high-level waste** – for example, used fuel rods.

Low-level waste can be sealed into containers and put in **landfill sites**.

Intermediate-level waste is mixed with concrete and **stored** in **stainless-steel containers**. It must be stored for thousands of years.

High-level waste is kept in **cooling tanks** at first because it decays so fast it gets hot. Eventually it becomes intermediate-level waste. High-level waste includes '**weapons grade plutonium**', which is the radioactive element plutonium, produced in nuclear reactors. It can be used to make nuclear weapons.

Waste can be **dispersed** (for example when sewage is discharged in the sea), or **contained** (for example when rubbish is put in a landfill site). Some radioactive waste is too dangerous to be dispersed.

Where to store the waste?
- At the bottom of the sea – but containers may leak.
- Underground – but containers may leak, and earthquakes or other changes to the rocks may occur.
- On the surface – but needs guarding (for example, from terrorists) for thousands of years.
- Blast into space – but danger of rocket explosion.

The **precautionary principle** says that if we are not sure of the effect of something, and it could be very harmful, then we should not risk trying it. (It is better to be safe than sorry.)

HOW SCIENCE WORKS

Exposure to radiation

In February 2006, a court found AEA Technology, a company specialising in the disposal of radioactive materials, guilty of exposing employees to radiation. The firm was fined £250 000. This is how the event might have been reported:

Toxic trail across the Pennines

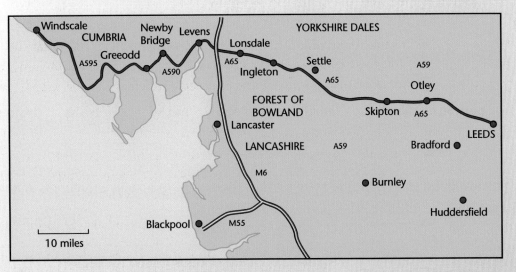

A container of radioactive waste emitting a lethal beam of radiation, was taken on a 130-mile journey across the Pennines. Fortunately, no-one was exposed to the beam during the journey from Cookridge hospital in Leeds to the Windscale waste reprocessing plant at Sallafield in Cumbria.

If anyone had stood about a metre from the 2.6 tonne container they would have exceeded the legal dose for employees in a few seconds, and suffered radiation burns within minutes. After two hours they would have been dead.

The cobalt 60 source was used for treating cancer patients, and had reached the end of its useful life. AEA Technology had the job of moving the source and taking it to Windscale to be safely stored. The staff arrived and put the cobalt 60 source in a container which had a shield plug missing from the under-side. As a result, a narrow beam of gamma rays was fired into the ground during the whole of the journey. The container was left in a locked area overnight. The next day high levels of background radiation in the area were recorded by a health physicist checking the area. The company are accused of using the wrong packaging, failing to notice the missing plug, and not training employees in the correct procedures.

Read the report carefully. What does it tell you?
- Although the source was very dangerous, the beam was pointing downwards, which is why no one was irradiated.
- The reason the hospital had such a dangerous source was because it was used to kill cancer cells.
- The container must have had thick lead walls as they stopped the gamma radiation.
- The journalist has not made it very clear whether a few seconds dose would result in death two hours later, or whether a two-hour dose would be needed.

Exam practice questions

1. Nuclear radiation passes through a sheet of aluminium foil. The radiation could be:
    (a)   alpha radiation only

    (b)   beta radiation only

    (c)   alpha, beta or gamma radiation

    (d)   beta or gamma radiation   **[1]**

2. Complete the following, using the words below:

    Radioactive emissions come from the _____ of _____ atoms.

    carbon   hydrogen   unstable   stable   nuclei   electrons   protons
                                                                                             [2]

3. Following the alara principle means that exposure to radioactive materials is kept
    __ _____ _____ _____ _____.   **[1]**

4. Which of the following is a significant source of background radiation?
    (a)   alpha radiation

    (b)   mobile phones

    (c)   rocks

    (d)   microwave ovens   **[1]**

5. A patient is given a dose of iodine-131 which has a half-life of eight days. What fraction of the iodine-131 nuclei are active after 32 days?
    (a)   1/4

    (b)   1/8

    (c)   1/16

    (d)   1/32   **[1]**

Exam practice questions

6. Use the words below to complete this description of a nuclear reactor (words may be used more than once or not at all):

The ____ ____ consist of the isotope of uranium–235. When a uranium ____ absorbs a ____ the nucleus splits into two smaller nuclei and a few more ____. A lot of energy is released. This is called nuclear ____. The neutrons produce a ____ reaction as they are absorbed by more uranium nuclei. The reaction is slowed or stopped by moving ____ ____ into the reactor. The ____ removes the heat energy from the reactor and uses it to heat water in the power station.   **[11]**

> **nucleus   fuel   neutron   control   chain   rods   coolant
> electron   atom   fission   protons   neutrons**

7. Complete the following using the words **high, low** or **intermediate**:
The used fuel rods from nuclear reactors are ____ level radioactive waste.
The protective clothing worn by radiation workers is ____ level radioactive waste.   **[2]**

8. (a) Which type of nuclear radiation is used in a paper thickness detector?
     (b) Explain why this type is chosen.   **[3]**

9. Describe how nuclear radiation is used to treat cancer.   **[2]**

10. (a) A patient is injected with technetium-99m which has a half-life of 6 hours. What fraction of the technetium-99m nuclei are left after one day?
     (b) Why is it better to use an isotope with a half-life of 6 hours rather than:
          (i)   an isotope with a half-life of 6 minutes,
          (ii) an isotope with a half-life of 6 days?   **[3]**

11. (a) What is nuclear fission?
     (b) How does the energy produced by nuclear fission compare with that produced in a chemical reaction?   **[2]**

12. Bob says that this high-level radioactive waste should be stored at the power station until a safe way to dispose of it is found. This is an example of the 'precautionary principle'. Explain what this means.   **[2]**

The following topics are covered in this chapter:

- **The effect of heating**
- **Heat transfer**
- **Energy resources**
- **How the electricity supply works**
- **Electrical resistance**

10.1 The effect of heating

Temperature and heat

OCR B   P1a

Temperature is a measure of how hot an object is. The **temperature scale** we use measures temperature in **degrees Celsius (°C)**. On this scale 0°C is defined as the temperature at which pure ice melts – but temperatures can be much lower than this. Different temperatures can be shown on a **thermogram** – each colour represents a different temperature.

> If the temperature changes *to* 1°C it is just above the freezing point of water – this is not the same as *by* 1°C. Don't confuse temperatures with temperature changes.

Heat is a form of energy. If an object is cooled until all the particles stop moving then they cannot lose any more kinetic energy. The amount of heat in the object is a minimum.

Fig. 10.1 A thermogram of an elephant.

Specific heat capacity

OCR B   P1a

If the temperature of a solid, liquid or gas changes, then it has gained, or lost, energy. The amount of energy depends on:
- the temperature change
- the mass of the object
- the material the object is made from.

The **specific heat capacity** of a material is a measure of the energy of the material. It is different for different materials and tells us how much energy (in joules) you need to raise the temperature of one kilogram of the material by one degree Celsius ('specific' means 'for each kilogram').

> **KEY POINT**
>
> The specific heat capacity of a material is the energy needed to increase the temperature of 1 kg of the material by 1°C:
>
> Energy = mass × specific heat capacity × temperature change
>
> Specific heat capacity is measured in J/kg °C.

Latent heat

OCR B   P1a

Heating an object raises its temperature except at the **melting point** and **boiling point**. At these temperatures, the energy is being used to **change the state,** from solid to liquid, or liquid to gas. Because the temperature does not change, the heat given to the substance is called **latent heat** ('latent' means hidden).

Solid objects are held together by forces between the particles (atoms or molecules), and have a regular shape. The particles vibrate more as the object is heated. **Liquid** particles have enough energy to break the inter-molecular bonds and slide over each other. At the **melting point**, heating the solid does not increase the vibrations, but gives the particles enough energy to break the bonds. When a liquid **freezes**, it loses this energy to its surroundings.

Gas particles have enough energy to separate completely. At the **boiling point**, heating the liquid breaks the inter-molecular bonds completely and the particles form a gas.

> A sketch of particles must show that the particle size doesn't change. Solid particles are regularly spaced and touching.

> The liquid particles are still touching – there are no gaps large enough for another particle to fit in.

> Gas particles are very widely spaced, so do not draw too many.

solid   liquid   gas

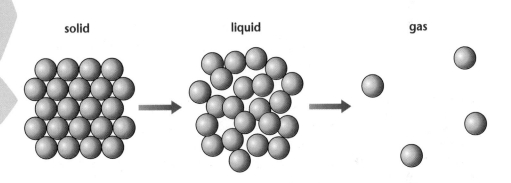

Fig. 10.2 There is a small increase in kinetic (movement) energy of the particles when an object changes from solid to liquid and a bigger increase when a liquid changes to gas.

KEY POINT

The specific latent heat of melting of a material is the energy in joules needed to melt 1 kg of the material without changing its temperature.

The specific latent heat of boiling of a material is the energy in joules needed to boil 1 kg of the material without changing its temperature. Specific latent heat is measured in J/kg.

The specific latent heat of freezing is the same as that for melting, but the energy is given out by the material.

10.2 Heat transfer

Heating and cooling

Objects that are hotter than their surroundings cool down, objects that are colder than the surroundings heat up. The bigger the temperature difference between an object and its surroundings, the faster this happens.

Heat transfer by conduction

In a hot solid, the atoms vibrate more than in a cold one. They collide with atoms next to them and set them vibrating more. The kinetic energy is transferred from atom to atom. Metals are the best **conductors**, followed by other solids. Liquids are generally poor conductors. Gases are very poor conductors. Poor conductors are called **insulators**.

Metals are good conductors because they have 'free' electrons that transport energy from the hot to the cold end of the material much faster.

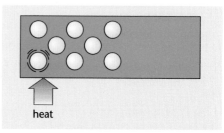

Fig. 10.3 Conduction in a solid. Energy is transferred from molecule to molecule.

Heat transfer by convection

> Remember: hot gases and liquids rise, not 'heat'.

In a hot fluid (a gas or a liquid), the atoms have more kinetic energy than in a cold fluid, so they move more. They spread out and the fluid becomes less dense. The hot fluid rises above the denser cold fluid forming a **convection current**.

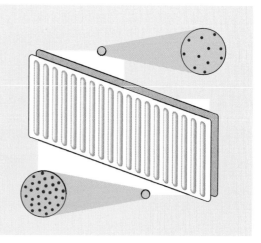

Fig. 10.4 The particles in the warm air are more widely spaced than those in the cold air, so the air is less dense.

Heat transfer by radiation

All objects **emit** (give out) and **absorb** (take in) **thermal radiation**. Some objects also **reflect** radiation. This radiation transfers energy in the form of infra-red electromagnetic waves. The hotter a body is the more energy it radiates. Radiation will travel through a vacuum – it does not need a **medium** (material) to pass through.

Dark and **matt** surfaces are **good absorbers** and **emitters** of radiation. **Light** and **shiny** surfaces are **poor absorbers** and **emitters** of radiation.

Insulation

In the winter we keep our homes much warmer than the outside temperature. This means that heat will be lost to the outside. If we reduce the heat lost, we use less fuel and it costs less. We can do this by **insulating** our homes.

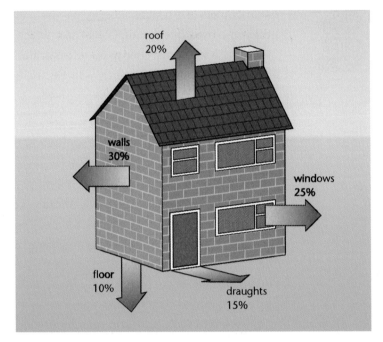

Fig. 10.5 Energy flow from an uninsulated house.

- **Still air** is a very good **insulator**. Some houses have **cavity walls**. The air gap between the two walls stops conduction. But air transfers heat if convection currents are set up. It is important to keep the air still. This can be done by **cavity wall insulation** – filling the cavity with a material containing trapped air, for example, foam or mineral wool.
- A lot of energy is lost through the roof, because **convection currents** are set up. **Loft insulation** uses fibreglass or mineral wool to keep the floor of the loft from getting hot.
- **Reflective foil** on walls reflects infra-red radiation.
- **Draught-proofing** stops the hot air leaving and cold air entering the house.

All these improvements cost money to buy and install, but they save money on fuel costs. You can work out the **payback time**, which is the time it takes before the money spent on improvements is balanced by the fuel savings, and you begin to save money.

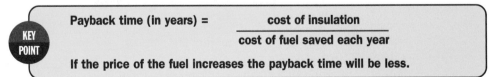

KEY POINT

Payback time (in years) = $\dfrac{\text{cost of insulation}}{\text{cost of fuel saved each year}}$

If the price of the fuel increases the payback time will be less.

10.3 Energy resources

Energy questions

OCR A | P3.3
OCR B | P2c
AQA | P1.13.4
EDEXCEL 360 | P1a10

As members of a modern society we use a lot of **energy resources**. Some of these are used directly, especially oil – which is used as petrol and diesel for vehicles. A lot of our energy resources are used to make **electricity**, which we then use as a source of energy – so electricity is called a **secondary energy source**.

Readily available energy has the **benefit** of making our lives healthier and longer. The **drawbacks** are the amount of damage to the environment – the **pollution**. When we choose a fuel, we need to consider the effect on the environment. Some waste is toxic. **Carbon dioxide** is a waste product from using many fuels. Increased levels of carbon dioxide in the atmosphere cause some **global warming** (see pp. 68 and 195). There is a lot of concern about climate change caused by global warming, so some countries, including the UK, are aiming to cut their emissions of carbon dioxide. The **risks** of using some energy resources are higher than others, but the risks to people because of extended power cuts are also high – for example; failure of hospital equipment, no heating in the winter and no refrigeration in the summer.

Efficiency of energy transfer

OCR A | P3.3
OCR B | P2b
AQA | P1.13.2
EDEXCEL 360 | P1a10

In a coal-fired power station for every 100 J of chemical energy stored in the coal that is burned, only 40 J is transferred to electrical energy. **Energy** is a **conserved** quantity, which means that the total amount of energy remains the same. It cannot be created. The 60 J of chemical energy that is not turned into electrical energy is wasted, and ends up as heat in the surrounding environment. This will cause a slight **temperature increase** in the surroundings. Energy becomes increasingly spread out and more difficult to use for further energy transformations.

Energy transfers can be shown on a **Sankey energy flow diagram**.

The width of the arrows is proportional to the amount of energy represented by the arrow.

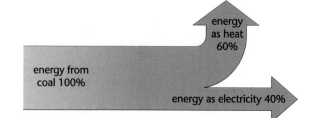

Fig. 10.6 Energy transformed in a coal-fired power station.

Modern gas-fired power stations are more **efficient** than coal-fired power stations. More of the chemical energy stored in the gas is transferred to electrical energy.

> KEY POINT
>
> The efficiency of an energy transfer is the fraction, or percentage, of the energy input that is transferred to useful energy output:
>
> Efficiency = useful energy output / total energy input
>
> Or as a percentage:
>
> Efficiency = useful energy output / total energy input × 100%

Gas-fired power stations are about 50% efficient whereas coal-fired power stations are about 40% efficient.

Generating electricity

OCR A | P3.3
OCR B | P2b
AQA | P1.13.4

Turning a **generator** produces electricity. To turn the generators we connect them to **turbines** and we use all the different energy resources available to turn the turbines. **Wind** and **water flow** can turn turbines directly. **Steam** is often used, produced by **heating water**. The heating is done by burning fuels, or using other heat sources. The diagram shows the parts of a coal-fired power station. In a **modern gas-fired** power station, the **hot exhaust gases** from the burners are used to turn the turbines, and then to heat water to steam which turns the turbines.

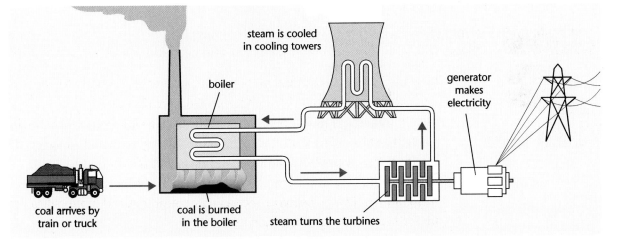

Fig. 10.7 A coal-fired power station.

The main fuels

OCR A | P3.3
OCR B | P2c
AQA | P1.13.4
EDEXCEL 360 | P1a10

Electricity can be generated in large power stations from many different types of fuel:
- **Fossil fuels** – coal, natural gas and oil. These fuels were formed from the remains of **forests 300 million years ago**. None is being formed today so they will eventually run out. When they are burned, **sulphur dioxide** gas is formed, which dissolves in rain to form **acid rain**. Coal produces more sulphur dioxide than oil, and gas produces less. All the fossil fuels produce **carbon dioxide** when burned.

- **Nuclear** fuels – the nuclei of **plutonium** and **uranium.** Uranium is mined, but will not run out as quickly as fossil fuels. Plutonium is formed in nuclear reactors. If the plutonium is not processed into fuel, some of it can be used to make nuclear bombs. Energy is released when **nuclear fission** occurs – the nucleus splits in two. One advantage is that no carbon dioxide is formed. The radioactive materials produced remain dangerous to living things for millions of years (see p. 152). These include **pollution** from the fuel processing (both producing the fuel and making the waste fuel safe after it is used) and from the reactor when it reaches the end of its life. In addition there is the risk of an accidental emission of **radioactive** material while the power station is operating. These concerns mean that there are high maintenance costs and high decommissioning costs (the costs of taking the plant apart and making it safe at the end of its life).

- **Renewable** fuels. These are covered in the next section. They are fuels that are being made today and so will not be used up. Most of these are not used in large power stations. An exception is **hydroelectric** power, generated in some countries from the flow of large rivers and high waterfalls.

The table shows the fuels we use in the UK for generating electricity.

Fuel used in 2004	Percentage
coal	40
oil	<1
gas	34
nuclear	24
hydroelectric	<1
other renewables	<1
imports (electric cable to France)	<1

The UK oil and gas reserves under the North Sea are being used up. Since 2004, the UK has been a net importer of gas (it imports more than it exports). The UK is expected to become a net importer of oil in 2010.

The renewable fuels

Most of these fuels make use of the **Sun's energy**. It is the Sun that evaporates water and causes the rain to fill the rivers. It also causes convection currents that produce winds. The exceptions are **geothermal,** which results from radioactive decay inside the Earth, and **tides,** which are caused by the Moon.

- **Hydroelectric power (HEP).** Fast-flowing water can be used to generate electricity. The UK does not have the fast-flowing large rivers needed to build large HEP stations, although there are some small ones in Scotland and Wales. There is no waste or pollution, but rainfall or snow is not constant so **dams** are needed. Building dams and flooding valleys and canyons changes the environment and causes conflict in some countries.

Fig.10.8 A wind farm - visual pollution?

- **Wind turbines**. These transfer the kinetic energy of the air into electrical energy. In 2006 there were about 1500 wind turbines in the UK, and the number is growing. Some areas, particularly offshore, are windy all year round. Wind turbines can be made **rugged** (tough) enough to last in these conditions. They do not produce polluting waste gases, but some people consider them noisy and an eyesore – **visual pollution**. The wind is unpredictable and the amount of electricity generated will depend on the wind speed. Wind turbines take up a lot of space for the amount of electricity generated.

- **Solar cells**. These are not used to turn generators. They produce electricity inside the cell. This means there is no need for overhead power cables. The electricity is **direct current (d.c.)** – the direction of the current does not change. There is no polluting waste and no need to buy fuel. There are no moving parts so they are rugged and do not need much maintenance. They have a long life. They cannot produce power at night or in poor weather and are usually used to charge batteries. The big advantage is that they can be used in remote locations where there are no power lines. More details are given in the section on solar energy (see p. 164).

- **Biomass** fuels, for example wood, straw, manure and household waste. These are products that are being formed today by plants or animals. Power stations need a steady supply. Suppliers need to be sure the power station will not close. The fuels can be burned directly, or fermented to produce **methane** gas. Using them will produce pollution – **carbon dioxide**, other gases and ashes. In the case of waste and manure, this pollution would be produced even if they were not used as fuel.

- **Geothermal**. There is a lot of heat below the Earth's crust. In places where the crust is thin and the heat is close to the surface, geothermal power stations can use this heat. Examples are in New Zealand and Iceland. There is no pollution, although the effect of extracting the heat may change the environment, and these areas experience earthquakes and volcanic action, which may damage a power station.

- **Tides**. In some places the change in the **height** of the water due to the tide is large enough to make it worth using to generate electricity. The tide does not depend on the weather. These areas are often important natural areas and holding back the water to run through turbines destroys the habitats of many birds and other animals.

Solar energy

OCR A — P3.3
OCR B — P2a
AQA — P1.13.4
EDEXCEL 360 — P1a10

The Sun can be used as an energy resource without using generators. To collect the most energy from the Sun the collector must track the path of the Sun. Some installations do this, but some are set at the best angle and collect fewer of the Sun's rays. There are three ways of using the Sun's energy:

- **Passive solar heating** for buildings. Glass is transparent to light and short wave infrared radiation, but reflects longer wave infrared radiation. When the Sun shines on glass windows the light and short wave infrared radiation will pass through and warm the objects inside. The warm objects emit longer-wave infrared radiation, but this cannot escape through the glass. This effect is called the **greenhouse effect** because it explains how plants are kept warm inside a glasshouse. The effect can be used in **solar panels** to warm water as shown in Figure 10.9. The water can be circulated in pipes around the house – the circulating water is often used to heat the water in the hot-water tank.

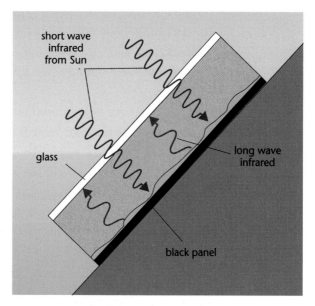

Fig. 10.9 A solar panel.

- **Solar furnace**. Light and infrared radiation is reflected from shiny surfaces. A curved mirror can be used to focus all the Sun's rays to a point. This can be used as a solar cooker to cook food.
- **Solar cells**. These are also called **photocells** or **photovoltaic cells**. They contain a crystal of **silicon**. Light falls on the crystal and gives **electrons** energy so they are released. The electrons flow as an **electric current**. The current can be increased by increasing the amount of light energy. This is done by increasing the surface area that the light falls on, or by increasing the light intensity. Solar cells are expensive to manufacture and are only about 30% efficient (although new developments may increase this to 50%). This is why they are not yet in widespread use.

10.4 How the electricity supply works

How generators work

OCR B · P2b
EDEXCEL 360 · P1.a9

The diagram shows how a voltage is **induced** in a coil of wire by **moving** a **magnet** into or out of a **coil**. Moving the coil instead of the magnet would have the same effect. This effect is used in **dynamos** and **generators**.

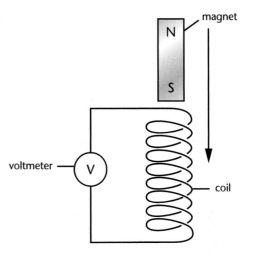

Fig. 10.10 A voltage is induced.

KEY POINT

The **dynamo effect** occurs when a **voltage is induced by:**
- moving a magnet near a coil
- moving a coil near a magnet.

There are three ways to increase the induced **voltage** (and get greater induced **current**):
- use stronger magnets
- use more turns of wire in the coil
- move the magnet (or the coil) faster.

The diagram shows a bicycle dynamo that uses a rotating magnet. As the magnet rotates faster, the induced current increases and the bicycle light gets brighter.

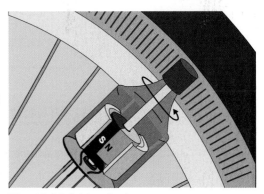

Fig. 10.11 A bicycle dynamo.

A generator in a power station uses an **electromagnet** to produce a **magnetic field**. The electromagnet rotates inside **coils of wire** so that the coil is in a **changing magnetic field** and a voltage is induced.

Alternating current (a.c.) and direct current (d.c.)

Changing the direction of the **magnetic field** or the **movement** induces a voltage in the *opposite* direction. As the magnet rotates, the north pole and south pole swap over once in each complete rotation. This means the direction of the voltage and the current changes. This is called **alternating current (a.c.)**.

> **KEY POINT** Dynamos and a.c. generators produce alternating current (a.c.). Batteries and solar cells produce direct current (d.c.).

The size of d.c. can change but it is always in the same direction. For a.c. a graph of the voltage alternates in the same way as a graph of the current.

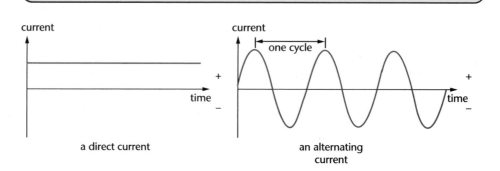

Fig. 10.12 d.c. and a.c.

In the UK, the a.c. generators at power stations that supply our mains electricity rotate **50 times in one second**. This means that there are 50 complete cycles each second. The number of cycles per second is called the **frequency**. Frequency is measured in cycles per second or in hertz (Hz), where 1 Hz = 1 cycle per second.

> **KEY POINT** in the UK, the mains frequency is 50 Hz, which is 50 cycles per second.

Electrical energy and power

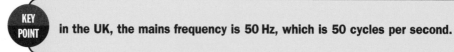

We use electrical appliances at home to transfer energy from the mains supply to:
● heating
● light
● movement (including sound).

Sound is made by the movement of a loudspeaker cone.

In two hours, an electric lamp transforms twice as much energy as it transforms in one hour. The **power** of an electrical appliance indicates how much **electrical energy** it transfers in one **second** – in other words, the rate at which it transfers electrical energy into other forms of energy..

> **KEY POINT** Electrical power is calculated using:
> Power = current × voltage
> $P = IV$
> Power is measured in watts (W) where 1 W = 1 J/s.

Appliances used for heating have a much higher power rating than those used to produce light or sound.

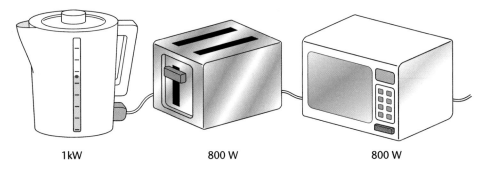

1 kW   800 W   800 W

Fig. 10.13 Power ratings of electrical appliances.

> **The kilowatt-hour is a unit of energy – not power. (Power is measured in watts or kilowatts.)**

The amount of **energy transferred** from the mains appliance depends on the **power** rating of the appliance and the **time** for which it is switched on. Energy is measured in joules, but electricity suppliers sell us electrical energy in **units** called **kilowatt-hours**. Electricity meters measure the energy transformed in kilowatt-hours.

KEY POINT

Electrical energy is calculated by

Energy = power × time

$$E = Pt$$

There are two sets of units used for energy:

- **The energy is in joules (J) when the power is in watts and the time is in seconds.**
- **The energy is in kilowatt-hours (kW h) when the power is in kilowatts and the time is in hours.**

The cost of each unit of electrical energy – which is one kilowatt-hour of electrical energy – varies. At the moment it is about 10p. The electrical energy bill is calculated by working out the number of units used and multiplying by the cost of a unit.

KEY POINT

Cost of electrical energy used is calculated from:

Cost = power in kW × time in hours × cost of one unit

or

Cost = number of kW h used × cost of one unit

For example:
For the 800 W microwave oven in Fig.10.13:
if it is used for half an hour and the cost of a unit is 10p:

Cost = 0.8 kW × 0.5 hours × 10p/kW h

Cost = 0.4p

Transformers and the National Grid

OCR B   P2b
AQA   P1.13.3
EDEXCEL 360   P1a10

Transformers

A **transformer** changes the size of an **alternating voltage**. The voltage must be changing or the transformer will not work. One of the reasons we use an a.c. mains supply is so that we can change voltage using transformers.

Step-up transformers increase voltage, and **step-down transformers** decrease voltage.

The National Grid

The National Grid is the network of suppliers of electricity – the power stations and users of electricity – homes and workplaces. They are all connected together by power lines, some overhead and some underground.

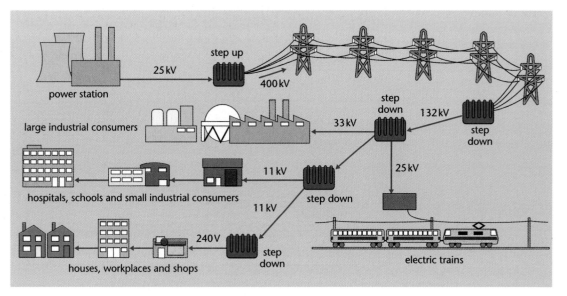

Fig. 10.14 The National Grid.

A National Grid has the following *advantages*:
- Power stations can be built where the fuel reserves are, or near the sea or rivers for cooling.
- Pollution can be kept away from cities.
- Power can be diverted to where it is needed, if there is high demand or a breakdown.
- Surplus power can be used to pump water up into reservoirs to be used to generate hydroelectric power when there is a peak in demand. (Dinorwig in Wales is a pumped storage power station.)
- Very large power stations can be built which are more efficient.

A National Grid has the following *disadvantages*:
- Power is wasted heating the power cables.
- Overhead power cables are an eyesore.
- Smaller generating projects such as wind turbines and panels of solar cells have difficulty competing with large suppliers.

We reduce the resistance of the cable by using thick copper, but the advantage of lower resistance has to be balanced against increased cost of cables and supports for the heavier cables.

Reducing power loss in power lines

To supply 100 kW of power through overhead power cables we could transmit 1 A at 100 kV or 10 A at 10 kV. (Using $P = IV$ the power is $P = 1A \times 100V = 100$ kW or $P = 10A \times 10V = 100$ kW.)

The power cables have a **resistance** to the flow of current. The heating effect in the cables depends on the **resistance** and on the **current** in the cables. By making the current as small as possible we can reduce the energy wasted as heat in the cables. The current can be small if the voltage is large.

> **KEY POINT**
>
> When supplying power through cables, a large voltage allows us to use a small current and this reduces energy waste by reducing the heating of the cables.

Transformers are used to **step-up** the voltage. Several different voltages are used, for example the supergrid is at 400 kV as you can see in Fig. 10.14. **Step-down** transformers reduce this for us to use, for example, 240 V in our homes.

10.5 Electrical resistance

Resistance and resistors

EDEXCEL 360   P1a9

The higher the **resistance** of a circuit the less **current** passes through it for a given **voltage**.

>
>
> **KEY POINT**
>
> Resistance = voltage/current
>
> Resistance is measured in ohms (Ω) where current is in amperes (A) and voltage in volts (V):
>
> $R = V/I$ and also $V = IR$ and $I = V/R$

A **variable resistor** connected in a circuit changes the current in a circuit by changing the resistance. This can be used to change how circuits work, for example to change how long the shutter is open in a digital camera.

In some components, such as resistors and metal conductors the resistance stays constant when the current and voltage change, providing that the temperature does not change. This means that the **current is proportional to the voltage**. A graph of current against voltage will be a straight line. **Ohm's Law** states that current is proportional to voltage for a conductor, as long as its temperature is kept constant.

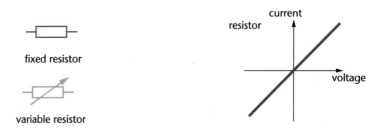

Fig. 10.15 A graph of current against voltage for a resistor.

Changing resistance

EDEXCEL 360 P1a9

The wire in a **filament lamp** gets hotter for larger currents. This increases the resistance so the graph of current against voltage is not a straight line.

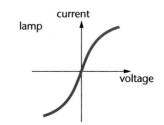

Fig. 10.16 A graph of current against voltage for a filament lamp.

The resistance of a **light-dependent resistor (LDR)** decreases as the intensity of light falling on it increases. This can be used in a circuit to control when a lamp switches on or off.

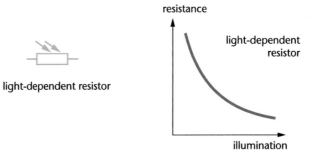

Fig. 10.17 A graph of resistance against intensity of light for a light-dependent resistor.

The resistance of a negative temperature coefficient (NTC) **thermistor** decreases as the temperature increases. This can be used to switch on a heating or cooling circuit at a certain temperature.

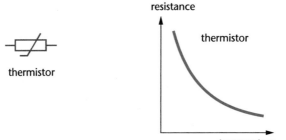

Fig. 10.18 A graph of resistance against temperature for a thermistor.

HOW SCIENCE WORKS

Nuclear reactors

In the UK, a number of nuclear reactors are reaching the end of their useful life. Our electricity requirements are expected to increase in the future. How should we generate our electricity in the future? There is no 'correct' answer. The government consults people and interested groups about their views and then decides. If people feel very strongly about the decisions they organise protests.

66 *Wind turbines don't cause pollution, and I don't think they are an eyesore – pylons and power stations look much worse.* 99

66 *Nuclear power stations produce dangerous waste that will be radioactive for millions of years – it is not fair to leave that for our descendents. Suppose there is an accident? The whole country may be dangerously radioactive.* 99

66 *The world could survive a nuclear power station blowing up – but global warming could make it uninhabitable – like Venus. We must stop burning fossil fuels.* 99

66 *The whole country would have to be covered with windmills to generate enough electricity. There will be no power when the wind drops – or in gales when it is too windy to use wind turbines safely.* 99

66 *We should think small – lots of solar cells and small wind turbines. Battery technology is improving all the time so we store the energy for when there is no sunshine or wind.* 99

66 *It's crazy to say we won't build nuclear power stations and then buy electricity from France – generated in their nuclear power stations. Modern nuclear power station designs are much safer than the old ones.* 99

Exam practice questions

1. Which of these improvements, to the insulation of a building, works by stopping convection:
   - **(a)** Two walls with a gap between (cavity walls)
   - **(b)** Two panes of glass with a gap between (double glazing)
   - **(c)** Filling the wall cavity with foam (cavity wall insulation)
   - **(d)** Silver foil wall lining behind radiators   **[1]**

2. A power station burns coal and converts 3.50 MJ of chemical energy to 1.19 MJ of electrical energy every second. The efficiency of the power station is:
   - **(a)** 2.31%
   - **(b)** 29%
   - **(c)** 34%
   - **(d)** 47%   **[1]**

3. When we 'use electricity' what do we measure in kilowatt-hours?
   - **(a)** cost of electricity supplied
   - **(b)** energy used
   - **(c)** power output
   - **(d)** time taken to use 1 kilowatt   **[1]**

4. Which of the following devices transfers kinetic energy to electrical energy?
   - **(a)** an a.c. generator
   - **(b)** an electric motor
   - **(c)** a turbine
   - **(d)** a transformer   **[1]**

In questions 5–9, fill in the gaps, using words once, more than once or not at all.

5. Water has a specific heat capacity of 4200 J/kg °C. This means that when one kilogram of water is given _____ J of energy its temperature will rise by 2°C. The specific heat capacity of aluminium is 880 J/kg °C. If one kilogram of aluminium is given 440 J of energy its temperature will rise by ___ °C.
   The heat needed to increase the temperature of a kilogram of water by 5°C is _____ the heat needed to increase the temperature of a kilogram of aluminium by 5°C.

   2100   4200   8400   0.5   1   2   more than   less than   the same as
                                                                                                         [3]

6. To improve the insulation of a building you need to reduce the heat lost by _____, _____ and _____.
   Loft insulation works by reducing _____ _____ in the roof space.
   _____ _____ work because the ____ ____ reduce conduction through the walls.
   Putting foil behind radiators reflects _____.

   air   cavity   conduction   convection   currents   gaps   radiation   walls   **[9]**

Exam practice questions

7. Most of the electricity in the UK is generated using _____, _____ and _____ fuel. To reduce the amount of _____ _____ gas emissions we want to reduce our use of _____ _____. Renewable fuels we could use include _____ _____ , but _____ of windmill farms would be needed to replace just one power station. We could build more _____ power stations, but these produce _____ _____that will remain _____ for _____ of years.

> **carbon dioxide   coal   fossil fuel   gas   hundreds   millions   nuclear radioactive   waste   wind energy**

[11]

8. A generator for mains electricity in the UK produces an _____ current which has a _____ of _____ cycles per second (Hz).

> **amplitude   alternating   direct   fifty   frequency   twenty**

[3]

9. A kilowatt-hour is a unit of _____. Electricity bills charge for each _____ used where one _____ is equal to a kilowatt-hour. If a unit costs 10p using a 3 kW electric fire for two hours will cost _____.

> **energy   joule   power   unit   15p   20p   30p   50p   60p**

[3]

10. **(a)** Why does a carpet feel warmer to bare feet than a stone floor?
   (b) Why are two thin blankets usually warmer than one thick one?
   (c) Why should you crawl to escape a smoke filled room?
   (d) Why is the heating element of an electric kettle in the bottom of the kettle?
   (e) Why is the back of a refrigerator painted black? **[5]**

11. The diagram shows a vacuum flask for keeping food hot, or cold.

air-filled stopper

double-walled
glass bottle

vacuum
between
glass walls

silvered
surfaces

(a) Explain how these features keep the food inside the flask hot:
   (i) the stopper
   (ii) the vacuum between the walls
   (iii) the silvering on the walls.
(b) Explain why the flask can be used to keep food cold, as well as hot. **[4]**

Exam practice questions

12. A combined cycle gas turbine (CCGT) power station burns gas and uses the hot exhaust gases to turn the turbines. The hot exhaust gases are then used to heat water to steam to turn more turbines. Every second 755 MJ of electricity is generated from 1300 MJ of chemical energy in the gas.

(a) Use the equation $\dfrac{\text{electrical energy output}}{\text{fuel energy input}} \times 100\%$

to calculate the efficiency of the power station. **[2]**

(b) Explain two ways a CCGT power station is different from a coal-fired power station. **[2]**

(c) Explain one disadvantage of using gas to fuel a power station. **[1]**

13. This table shows some electrical power ratings for wind generators.

Model	Maximum power rating of one turbine (kW)	Wind turbine recommended for:
A	0.025	battery charger
B	0.5	electricity for a caravan
C	5	domestic electricity for a house
D	25	electricity for a school
E	500	a wind farm
F	1500	an offshore wind farm

(a) What will affect the amount of power that a wind turbine can generate? **[1]**

(b) What model is recommended for a house? **[1]**

(c) A house has an average power use of 0.5 kW. Describe how the actual power used might be **(i)** higher than average and **(ii)** lower than average **[2]**

(d) Suggest two reasons model B is not recommended for a house. **[2]**

(e) Wind turbines are often used to recharge banks of batteries. Why is this a good idea? **[1]**

(f) A power station generates 750 MW of electrical power. How many wind turbines would be needed to replace it with **(i)** an offshore wind farm and **(ii)** an onshore wind farm **[2]**

(g) What is meant by 'the pay-back time?' **[1]**

Exam practice questions

14. Ian decides to check his electricity bill:

Meter reading at start of quarter	487 612 kilowatt-hours
Meter reading and end of quarter	489 360 kilowatt-hours
Cost of one unit	10p

(a) How many units has Ian used? **[1]**

(b) How much is his electricity bill for this number of units? **[1]**

Ian decides to try to cut down the amount of electricity he is using so he keeps some notes for a week:

Appliance	Power rating (kW)	Time used each day (hours)
electric fire	3	2
electric kettle	1.2	0.5
electric lamp	0.1	5
computer	0.3	4
television	0.5	3

(c) How many units does the electric fire use in a day? **[1]**

(d) How much does it cost to use the fire for a day? **[1]**

(e) Which appliance is used for the most time? **[1]**

(f) How much does it cost to use this appliance for a day? **[1]**

The following topics are covered in this chapter:

- *Describing waves*
- *Seismic waves*
- *Sound and ultrasound*
- *Electromagnetic waves*
- *Wave properties*

11.1 Describing waves

Types of wave

A **wave** is a **vibration** or disturbance which is transmitted through a material – called a **medium** – or through space. Waves transfer **energy** and can also be used to transfer **information** from one place to another, but they do not transfer material.

> A common mistake is to mark the amplitude from the top of a peak to the bottom of a trough – this is twice the amplitude.

Transverse waves

A **transverse wave** has the vibrations at **right angles** (perpendicular) to the direction of wave travel. The wave has **peaks** (or **crests**) and **troughs**, as shown in the diagram. The **amplitude** is the maximum displacement (change in position) from the undisturbed position. The **wavelength** (symbol λ) is the distance between two neighbouring peaks or troughs.

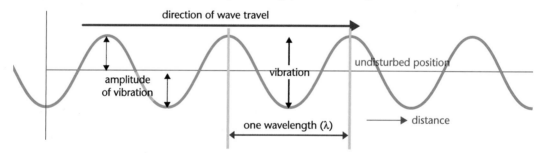

Fig. 11.1 A transverse wave.

The **frequency** is the number of complete waves that pass through a point in one second. It depends on how fast the source of the waves is vibrating. The frequency is measured in **hertz** (Hz) where one hertz is one cycle (wave) per second.

The **wave speed** depends on the medium that the wave is travelling through. As the frequency increases the wave speed does not change, but the wavelength will decrease. This is shown in Figure 11.2.

a wave on a rope   a higher frequency wave

Fig. 11.2 Waves with different frequency and wavelength.

The **wave equation** relates the wavelength and frequency to the wave speed.

> **KEY POINT**
>
> **The wave equation, for all waves:**
>
> **wave speed = frequency × wavelength**
>
> $v = f\lambda$
>
> **If f is in Hz and λ is in m, then v is in m/s.**
>
> **The equation can also be written as $\lambda = v/f$ and $f = v/\lambda$.**

Examples of transverse waves are: water waves, light and other electromagnetic waves, and the seismic waves called S-waves.

Longitudinal waves

A **longitudinal wave** has the vibrations **parallel** to (along the same direction as) the direction of the wave travel. As shown in the diagram, the wave has **compressions** (or squashed parts) and between these are stretched parts called **rarefactions**.

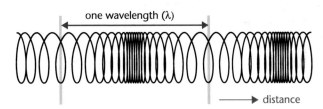

one wavelength (λ)

distance

Fig. 11.3 A longitudinal wave.

One wavelength is the length of one complete wave – a compression and a rarefaction. Longitudinal waves show the same behaviour (for example reflection and refraction) as transverse waves.

11.2 Seismic waves

P-waves and S-waves

OCR B   P1h
EDEXCEL 360   P1b11

There are two types of shock waves called **seismic waves** that travel *through* the Earth (other types travel over the surface.) These are called **P-waves** and **S-waves**. They can be caused by an earthquake, or a large explosion, and are detected at monitoring stations around the Earth by instruments called **seismometers**. Fig. 11.4 shows a **seismograph** – a record of the waves received at a monitoring station.

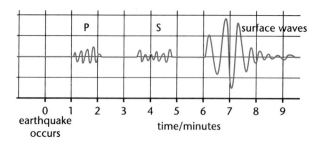

Fig. 11.4 A seismograph.

P-waves

P-waves, or primary waves, are **longitudinal** waves that travel through solid and liquid rock. They travel faster through the Earth than other seismic waves so they are the first to be detected after an earthquake.

S-waves

> Transverse waves can travel on the surface of liquids – but not through them.

S-waves, or secondary waves, are **transverse** waves and can only travel through the solid materials in the Earth. They are detected after primary waves because they have a lower speed.

Figure 11.5 shows the paths of P-waves and S-waves through the Earth following an earthquake. The point on the Earth's surface directly above the earthquake is called the **epicentre**. On the opposite side of the Earth to the epicentre no S-waves are detected. This tells us that there must be a part of the Earth that is liquid and only P-waves pass through it. This is how we know there is a **liquid outer core**.

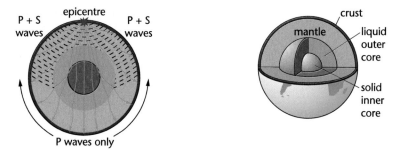

Fig. 11.5 P and S waves travel through the Earth.   **Fig. 11.6** The Earth's structure.

> **KEY POINT**
> The S-waves form a shadow region on the opposite side of the Earth to an earthquake. This shows that part of the Earth's core is liquid.

We can investigate the structure of the Earth's crust by setting up monitoring equipment at different points and setting off a controlled explosion. We record the waves arriving at the monitoring points. The time after the explosion that waves take to arrive depends on the speed of the waves (which in turn depends on the medium they pass through) and whether they have been reflected at a boundary between different materials. Analysing this data gives us information about the structure of the rocks.

11.3 Sound and ultrasound

Longitudinal waves

EDEXCEL 360   P1b11

Sound waves are **longitudinal** waves. They pass through solids, liquids and gases. **Ultrasound** waves are sound waves with a **frequency** that is too high for humans to hear. Like all waves, sound waves and ultrasound waves can be reflected – the reflections are called **echoes**. Echoes can be used for measuring distances. Figure 11.7 shows how the depth of water can be measured by reflecting an ultrasound pulse off the seabed.

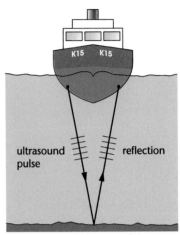

Fig. 11.7 Using echoes to measure distance.

Echo sounding works like this:
- An ultrasound pulse is emitted from a vibrating crystal.
- The same crystal detects the reflected pulse.
- The time for the pulse to travel from the crystal to the seabed and back to the crystal is recorded.
- The distance can be worked out using the equation speed = distance/time. The speed of the ultrasound waves in water is known, so distance = speed × time. The depth is half the distance the wave travelled,
so depth = $\dfrac{\text{speed} \times \text{time}}{2}$

> **When you work out the depth, remember that the pulse has travelled twice the depth – there and back.**

Ultrasound is also used to **scan** parts of the body, like the eye or an unborn foetus. This works because part of the pulse is reflected at each **boundary between** different **tissues** (for example skin and bone). The reflections from the tissue boundaries are all used to build up a picture. Ultrasound is much safer than X-rays because it does not damage body cells or DNA and does not cause mutations.

11.4 Electromagnetic waves

Transverse waves

OCR B   P1g
AQA   P1.13.5
EDEXCEL 360   P1b11

Electromagnetic waves are transverse waves, which are made up of vibrating magnetic and electric fields. They can travel through a **vacuum** and all travel through space at a speed of **300 000 km/s**. The different types of electromagnetic waves form the **electromagnetic spectrum**, shown in Figure 11.8.

frequency/Hz	10^{20}	10^{17}	10^{14}	10^{11}	10^{8}	10^{5}
	gamma rays		ultraviolet	infrared		radio waves
		X-rays		light	microwaves	
wavelength/m	10^{-12}	10^{-9}	10^{-6}	10^{-3}	1	10^{3}

Fig. 11.8 The electromagnetic spectrum.

The spectrum of electromagnetic waves is continuous from the **longest** wavelengths (**radio waves**) through to the **shortest** wavelengths (**gamma rays**). These **wavelengths** are related to the **frequencies** using the wave equation – radio waves have the lowest frequency and gamma rays the greatest frequency.

The **higher** the **frequency** the more **energy** the waves have. The wavelength of visible light is about half a thousandth of a millimetre – so small that it is not obvious that it is a wave. In fact, sometimes light behaves as a stream of particles and sometimes as a wave.

11.5 Wave properties

Absorption, emission and reflection

OCR B   P1d
AQA   P.1.13.5
EDEXCEL 360   P1b11

All surfaces **emit** electromagnetic radiation that depends on their **temperature**. They also **transmit**, **absorb** and/or **reflect** some of the radiation that falls on them. How much is transmitted, how much absorbed and how much reflected depends on the surface. Reflection happens to waves and particles. If a wave, or a ball, strikes a wall at an angle it will be reflected so that the **angle of incidence** is equal to the **angle of reflection**.

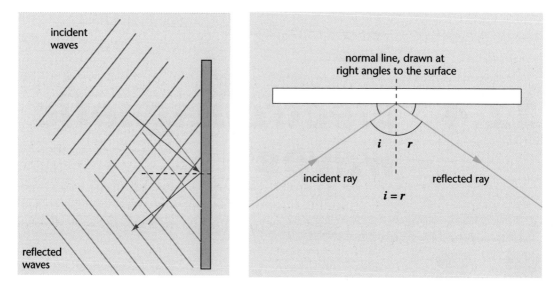

Fig. 11.8 Reflection.

Refraction

OCR B — P1e
AQA — P1.13.5
EDEXCEL 360 — P1b11

When waves enter a **denser medium**, they slow down. When they enter a less dense medium, they speed up. In both cases, this may cause them to **change direction**. This happens to water waves but also to particles, as you can show by rolling a ball at an angle down a ramp. When the slope of the ramp changes the direction of the ball changes.

> Water waves slow down as they go from deep water to shallow water.

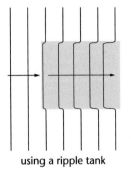
using a ripple tank

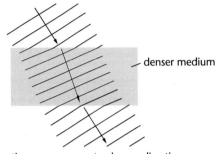
refraction causes waves to change direction

Fig. 11.10 Refraction.

Diffraction

OCR B — P1f
AQA — P1.13.5
EDEXCEL 360 — P1b11

Diffraction is the **spreading** out of a wave when it passes through a gap. The effect is most noticeable when the gap is the same size as the wavelength. Particles cannot be diffracted. So diffraction is good evidence for a wave.

> When answering questions it is important to say that the amount of spreading depends on the size of the gap compared with the wavelength

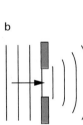 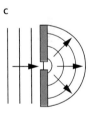

Fig. 11.11 Diffraction.

Interference

OCR B — P1f
AQA — P1.13.5
EDEXCEL 360 — P1b11

Interference occurs where two waves overlap. If the waves have the same **amplitude** and **wavelength** and are in **phase** (in step), they can **interfere**. Figure 11.12 shows that if the crests arrive **together** there will be **constructive** interference and the amplitude will **increase**. If a crest arrives at the same time as a trough the two will **cancel** out and there is no wave. This is **destructive** interference. Particles cannot interfere, so interference is good evidence for a wave.

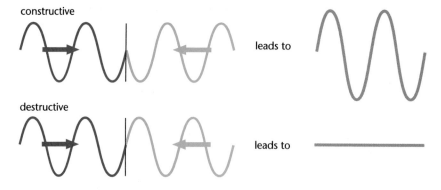

constructive

destructive

leads to

leads to

Fig. 11.12 Wave interference.

Total internal reflection

OCR B   P1e
AQA   P1.13.5
EDEXCEL 360   P1b11

Total internal reflection can only happen when light travels from a **dense to a less dense medium** – for example from glass to air. If the angle of incidence is so large that the angle of refraction would be greater then 90° then it is impossible for the light to leave the glass – so total internal reflection occurs.

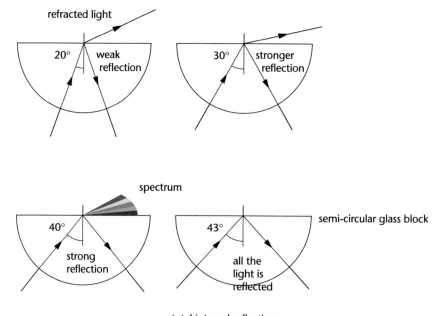

Fig. 11.13 When air meets an air–glass boundary.

Transparent materials have a **critical angle**. When the angle of incidence equals the critical angle, the angle of refraction is 90°. At angles greater than the critical angle, total internal reflection occurs. The critical angle for a glass–air boundary is about 42°. Total internal reflection can also occur at a Perspex–air boundary and at a water–air boundary.

HOW SCIENCE WORKS

Can earthquakes be predicted?

Each year on Earth, there are about 19 large earthquakes of magnitude 7.0 or higher. (Magnitude is a scale for measuring earthquake size.) This is an average and the table shows how the number has ranged from 6 large earthquakes in 1986 to 34 in 1957.

Most large earthquakes occur along the fault zones bordering the Pacific Ocean, so we can predict that this is *where* large earthquakes will occur, but it is not so easy to predict *when*. Some of the faults are well known and predictions are made of where and when an earthquake will occur based on how much time there has been between large earthquakes in the past. This forecasting technique can be used only for well-understood faults, such as the San Andreas fault in California. It is not possible for many faults, for example those that caused the 1994 earthquake in Northridge, California, and the 1995 earthquake in Kobe, Japan.

One successful earthquake prediction was for the 1975 Haicheng earthquake in China, when an evacuation warning was issued the day before the earthquake. People were warned to evacuate because there was an increase in the small earthquakes that can happen before a large one. Unfortunately, most earthquakes do not have these small warning earthquakes, for example there was no warning of the 1976 Tangshan earthquake in China, which caused an estimated 250 000 fatalities.

Large earthquakes (magnitude greater than 7) each year since 1956:

Year	Count	Year	Count
1956	15	1982	10
1957	34	1983	15
1958	10	1984	8
1959	15	1985	15
1960	22	1986	6
1961	18	1987	11
1962	15	1988	8
1963	20	1989	7
1964	15	1990	18
1965	22	1991	16
1966	19	1992	13
1967	16	1993	12
1968	30	1994	13
1969	27	1995	20
1970	29	1996	15
1971	23	1997	16
1972	20	1998	12
1973	16	1999	18
1974	21	2000	15
1975	21	2001	16
1976	25	2002	13
1977	16	2003	15
1978	18	2004	16
1979	15	2005	11
1980	18	Total 1900–2005 = 2061 earthquakes	
1981	14	Average: 19.4 each year	

HOW SCIENCE WORKS

The data shows that there is no pattern to the number of large earthquakes. It tells us that we can expect about 19 or 20 each year, but there have been years when there were as few as 6 and as many as 36, so this could happen again.

Some people claim to sense when an earthquake is about to happen. Others say that animals know and behave strangely. Scientists have studied these effects and found that people are just as likely to predict earthquakes that don't happen, and that animals behave strangely when no earthquake occurs.

How can we prevent loss of life in earthquakes?

It is not earthquakes that kill people, but collapsing buildings. In the 1920s, the geologist Bailey Willis tried to persuade Californians to pass safe building regulations that would make buildings less likely to collapse in an earthquake. He thought that the authorities were acting too slowly and began to exaggerate, talking of a serious earthquake within 10 years. A building organisation showed that his claims were scientifically unsound, and because he was discredited Willis had to give up his work and the building regulations were rejected. Then in March 1933 there was an earthquake at Long Beach in California in which 15 school buildings collapsed and 40 were damaged. Only the fact that the earthquake occurred at 5.45 pm prevented the deaths of thousands of children. This time scientists campaigned for safer building regulations with success and the regulations were passed in May 1933.

This example shows that it is important to use scientifically correct data and facts to support your arguments – otherwise people can point out scientific errors and use them to discredit the argument. Safe building regulations were a good idea, whether an earthquake occurred in 10 years or 50 years, but by saying that an earthquake would occur in 10 years, Willis lost the argument.

Since those first building regulations, the designs have improved and when there is an earthquake today, the earthquake engineers investigate immediately to see what can be learned and how building techniques can be improved for the future. Another way to save lives is to use the delay between the P-waves (which arrive first) and the more damaging S-waves to give a warning. The Japanese bullet train, the Shinkansen, has sensors to automatically detect the P-waves and shut down, so that at least the train is slowing down before the S-waves arrive. Sensors can be used to shut down other processes in a similar way.

Exam practice questions

1. Which word describes the number of waves passing a point in one second?
   (a) amplitude
   (b) frequency
   (c) wavelength
   (d) wave speed   [1]

2. Earthquake waves known as P-waves can pass through
   (a) liquids only
   (b) solids and liquids
   (c) solids only
   (d) neither solids nor liquids   [1]

3. The diagram shows how water waves spread out after passing through a gap.

   This effect is called
   (a) diffraction
   (b) dispersion
   (c) reflection
   (d) refraction   [1]

4. Use some of the words below to complete the sentences.
   A wave is the movement of a _____ through a _____. Waves transfer _____ but
   not _____. A wave is caused by something that _____.   [5]

   disturbance   energy   longitudinal   matter   medium   transverse   vibrates

5. Use some of the words below to complete the sentences.
   Waves change _____ at the boundary between two materials. The _____ also
   changes (the _____ does not change) and this may cause a change in _____
   of the wave.   [4]

   direction   frequency   shape   speed   wavelength

6. (a) Draw a diagram of a transverse wave.
   (b) Mark on the wavelength.
   (c) Mark on the amplitude.   [3]

7. An ocean wave has a wavelength of 200 m and frequency of 10 Hz.
   (a) How many complete waves are there for a time period of 1 s?
   (b) Write down the equation you can use to work out wave speed from wavelength and
       frequency.
   (c) What is the speed of these waves?   [3]

12 Electromagnetic radiation

The following topics are covered in this chapter:

- ***Energy and intensity***
- ***Wave communications***
- ***Radio waves and microwaves***
- ***Infrared and light***
- ***Ultraviolet, X-rays and gamma rays***
- ***The atmosphere and global warming***

12.1 Energy and intensity

The intensity of radiation

OCR A P2.1

Electromagnetic radiation carries **energy**. When a beam of radiation strikes a surface, the **intensity** of the radiation is the energy arriving on a surface area of one square metre each second.

> **KEY POINT** **The intensity of electromagnetic radiation is the energy arriving at the surface each second on a surface area of one square metre.**

The radiation is **emitted** from a **source** and travels towards a destination. On this **journey** the radiation **spreads out**, so the **further away** a **detector** is from the source, the less energy is detected. This is shown in Figure 12.1 The **intensity** of the radiation can be increased by moving closer to the source.

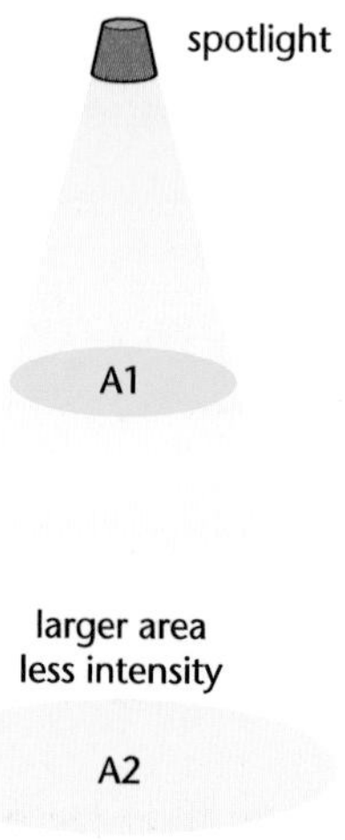

spotlight
A1
larger area
less intensity
A2

Fig. 12.1 Radiation spreads out from a source.

Area A2 is larger than A1, so the intensity is less. Intensity can also be increased by increasing the power (or total energy) of radiation from the source. One way of doing this is shown in the Fig.12.2.

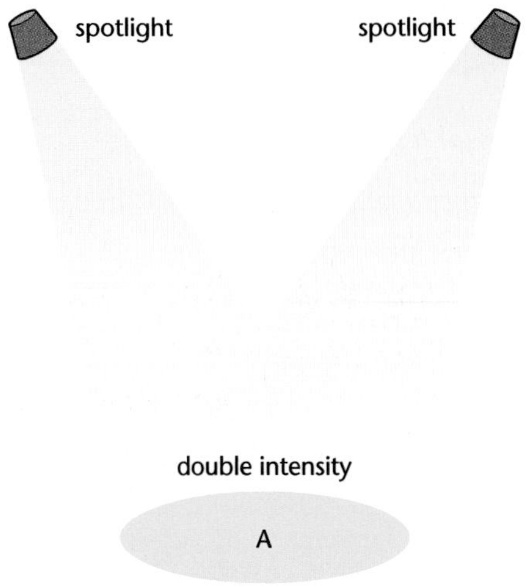

Some materials **absorb** some types of electromagnetic radiation. The **further** the radiation travels through an absorbent material, the lower its **intensity** will be when it reaches the end of its journey. This is because more of the radiation energy is absorbed by the material. The material gains energy from the incident radiation and heats up.

double intensity

A

Fig. 12.2 The identical spotlights on the same area double the intensity.

On some journeys, electromagnetic radiation crosses a boundary between two different materials and is **reflected**.

The energy of radiation

OCR A   P2.1

The **spectrum** of electromagnetic radiation ranges continuously from **low-energy radio waves** to very **high-energy gamma rays**. The **electromagnetic radiation** comes in packets of **energy** called **photons**. The table lists the types of electromagnetic radiation in order of increasing photon energy.

Types of electromagnetic radiation

lowest photon energy	radio waves
	microwaves
	infrared
	light spectrum — red / violet
	ultraviolet
	X-rays
highest photon energy	gamma rays

The boundary between each type of radiation is not a set value. For example, the highest-energy microwave photons merge into the lowest-energy infrared photons.

> **KEY POINT**
>
> Electromagnetic radiation delivers energy in 'packets' called photons.

The **energy** arriving at a surface will depend on the **number of photons** striking the surface and the **energy of each photon**. For example, a photon of blue light has more energy than a photon of red light, so in Figure 12.3, if

Gamma ray photon energies and X-ray photon energies overlap. A gamma ray photon comes from the nucleus of an atom and an X-ray photon does not, but they can have the same energy.

the number of red photons and the number of blue photons striking the area A is the same, then the blue spotlight illuminates the area A in the diagram with more energy than the red one.

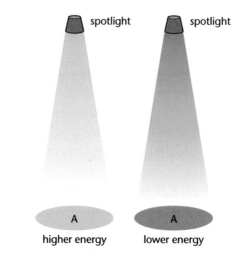

Fig. 12.3 Blue light has more energy than red light.

12.2 Wave Communications

Digital and analogue signals

OCR B P1e
AQA P1.13.5
EDEXCEL 360 P1b11

Electromagnetic radiation is used for **communications** and **transmission of information**. The waves that are used in this way are **radio waves**, **microwaves**, **infrared radiation** and **light**.

The idea of using a signal lamp to communicate was used in the 19th century. This method of long distance communication needed a code. One code used was **Morse code**, a series of long and short flashes of light for different letters of the alphabet. These signals can only be seen when visibility is good and for short distances.

Today we still use codes to send signals using electromagnetic radiation. There are two types of signal, **analogue** and **digital**. An analogue signal changes in frequency and amplitude all the time in a way that matches the changes in the voice or music being transmitted. A digital signal has just two values – which we can represent as **0 and 1**.

> **KEY POINT** An analogue signal varies in frequency and amplitude. A digital signal has two values, 0 and 1 (or 'on' and 'off').

The signal (voice, music or data) is converted into a code using only the values 0 and 1. The signal becomes a stream of 0 and 1 values. These pulses are added to the electromagnetic wave and transmitted. The signal is received and then decoded to recover the original signal.

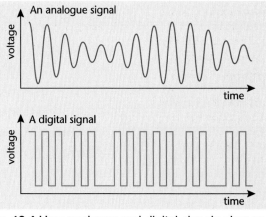

An analogue signal

A digital signal

Fig. 12.4 How analogue and digital signals change with time.

Both analogue and digital signals can pick up unwanted signals that distort the original signal. These **unwanted signals** are called **noise**. Digital signals can be cleaned up in a process known as **regeneration** because each pulse must be a 0 or a 1, so other values can be removed. Analogue signals can be **amplified**, but the noise is amplified too. This is why digital signals give a better-quality reception.

> **KEY POINT**
> Digital signals give a better-quality reception because noise on digital signals is more easily removed.

12.3 Radio waves and microwaves

Radio waves

OCR B   P1f
AQA   P1.13.5

A common mistake is to think that we can hear radio waves. We cannot hear any electromagnetic radiation. The radiation is used to carry a signal that is converted into a sound wave by the receiver.

Radio waves are the lowest-energy, lowest-frequency and longest-wavelength electromagnetic waves. They are produced when an **alternating current** flows in an **aerial** and they spread out and travel through the atmosphere. Another aerial is used as a detector and the waves produce an alternating current in it, with a frequency that matches that of the radio waves. Anyone with a receiver can tune it to this frequency to pick up the **radio waves** so they are suitable for **broadcasting** to large numbers of people. An advantage is that this method of communicating does not require wires to transmit information. A disadvantage is that radio stations using similar transmission frequencies sometimes **interfere**.

Medium wavelength radio waves are **reflected** from the **ionosphere**, a layer of charged particles in the upper atmosphere, so they can be used for long distance communication.

Digital radio has better-quality reception as it uses digital signals and so does not have problems of noise and interference.

Microwaves

OCR A	P2.1/2
OCR B	P1d/f
AQA	P1.13.5
EDEXCEL 360	P1b11

For OCR A you need to know about the energy of electromagnetic radiation. For the other specifications you need to know about the energy, frequency and wavelength of the electromagnetic waves.

Microwaves are sometimes considered to be **very short radio waves** (high-frequency and high-energy radio waves).

Some important properties of microwaves are:

- They are **reflected** by **metal** surfaces.
- They **heat materials** if they can make **atoms** or **molecules** in the material **vibrate**. The amount of heating depends on the **intensity** of the microwave radiation, and the **time** that the material is exposed to the radiation.
- They pass through **glass and plastics**.
- They pass through the **atmosphere**.
- They pass through the **ionosphere** without being reflected.
- They are **absorbed by water molecules**, how well depends on the frequency (energy) of the microwaves.
- Transmission is affected by wave effects such as reflection, refraction, diffraction and interference.

Microwaves and water molecules

A **microwave** frequency (energy) can be selected which is strongly absorbed by **water molecules**, causing them to **vibrate**, and increasing their **kinetic energy**. This effect can be used to heat materials containing water, for example **food**. If the most strongly absorbed frequency (energy) is used in a **microwave oven** it only cooks the outside of the food because it is all absorbed before it penetrates the food. So the frequency (energy) used in a microwave oven is changed slightly to one that will **penetrate** about **1 cm** into the food. Conduction and convection processes then spread the heat through the food. As our **bodies** contain water molecules in our **cells**, microwave oven radiation will heat up our cells and is very **dangerous at high intensity** because it will burn body tissue. The radiation is kept inside the oven by the **reflecting metal case** and **metal grid** in the door.

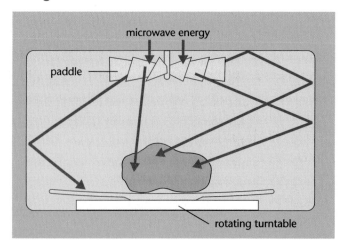

Fig. 12.5 A microwave oven.

Microwaves sent through the **atmosphere** will be absorbed by water so they can be used to **monitor rain**. The weaker the signal reaching the detector, the more rain the microwaves have passed through.

Microwave transmissions

Wireless technology uses **microwaves** and **radio waves** to transmit **information**. Advantages are:

- we can receive phone calls and email 24 hours a day
- no wiring is needed to connect laptops to the Internet, or for mobile phones or radio
- communication with wireless technology is portable and convenient.

Microwaves can be used to **transmit signals** over large distances if there are no obstacles between to reflect or absorb the beam. Another way to say this is that the transmitter and receiver are in **line of sight** (one can be seen from the other). This is why the transmitters are positioned high up, often on tall microwave masts. They cannot be spaced so far apart that, for example, hills or the curvature of the Earth stop the beam.

Microwaves are used to send signals to and from **satellites**. The satellites can relay signals around the Earth. Microwaves are used because they pass through the atmosphere and through the ionosphere. The signals may be for television programmes, telephone conversations, or monitoring the Earth (for example, weather forecasting).

There is more about diffraction on page 181.

When microwaves are transmitted from a dish the wavelength must be small compared to the dish diameter to reduce diffraction – the spreading out of the beam.

Mobile phones use microwave signals. The signals from the transmitting phones reflect off metal surfaces and walls to communicate with the nearest transmitter mast. There is a network of transmitter masts to relay the signals on to the nearest mast to the receiving phone.

Mobile phones have not been in widespread use for many years, so there is not much data about the possible **dangers** of using them. The transmitter is held close to the user's head so the microwaves must have a small heating effect on the brain. There are questions about whether this could be dangerous, or whether it is not large enough to be a problem. So far studies have not found that users have suffered any serious ill effects. There may also be a risk to residents living close to mobile phone masts.

KEY POINT Low-intensity microwave radiation, from mobile phone masts and handsets, may be a health risk, but there is disagreement about this.

12.4 Infrared and light

Infrared radiation

There is more about absorption and reflection on page 184.

When **infrared** radiation strikes our **skin**, we feel **heat**. Infrared radiation is **absorbed** by **black** and **dull** surfaces and is **reflected** from **silver** and **shiny** surfaces. When it is absorbed, all the particles in the surface are heated. So infrared radiation is used for **cooking** the surface of food (the interior is then heated by convection and conduction). For the same reason we must be careful that intense infrared radiation does not burn our skin.

Infrared radiation is used in remote controls for televisions and other electronic appliances such as DVD and video recorders. Looking with a digital camera (which shows up infrared signals that our eyes cannot see) you can see the flashing infrared-emitting diode sending the signal. These signals cannot pass through solid objects but can reflect off walls and ceilings to operate the television.

Communications using infrared radiation and light

There is more about internal reflection on page 182.

Infrared radiation and light travel along glass optical fibres by being totally internally reflected. The fibres are made with a core that has a different refractive index from that of the outer rim. The signal is reflected at the boundary as shown in Figure 12.6.

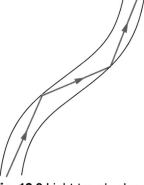

Signals (pulses of radiation) can be sent for long distances using **optical fibres**. A stream of data can be transmitted very quickly. There is **less interference** than with microwaves passing through the atmosphere. It is also possible to use

Fig. 12.6 Light travels along optical fibres.

multiplexing, a way of sending many different signals down one fibre at the same time. Digital signals are used so that noise can be removed when the signal is regenerated.

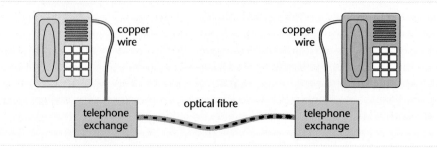
copper wire   copper wire

telephone exchange   optical fibre   telephone exchange

Fig. 12.7 Modern phones use optical fibres.

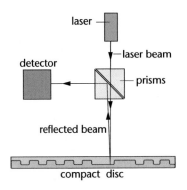

Lasers

Lasers produce a beam of radiation in which all the waves have the **same frequency** and are **in phase** (in step) with each other. Both infrared radiation and red light lasers are used as sources for fibre optic communications. They are also used in a **CD player.** Compact discs (CDs) store information digitally as a series of pits and bumps on its shiny surface. A laser beam is reflected differently from the pits and bumps and a detector is used to 'read' the different reflections as 0s and 1s – the information stored on the CD.

Fig. 12.8 A laser is used to play a CD.

Iris recognition

Iris scanning is increasingly being used to **identify** people when high security is needed. The iris is so **individual** that the chance of mistaken identity is almost zero. A **digital camera** is used to take a clear high-contrast picture of the iris using both **infrared** radiation and **light**. A computer is used to locate the pupil (which appears very black in infrared illumination). The **patterns in the iris** are translated into a code. Two hundred points are compared in the iris (only 60 or 70 are used when matching fingerprints). The computer looks for features such as rings and freckles. The process takes about 5 seconds. The iris does not change much over a lifetime, and the technique works even when people wear contact lenses.

12.5 Ultraviolet, X-rays and gamma rays

Ionising radiation

Ultraviolet radiation, X-rays and gamma rays are **high-energy** radiations that can **ionise** atoms they hit. If these atoms are ones inside the body the radiation can damage or kill cells. If the **DNA** in a **cell** is **damaged** it may **mutate**. This can cause cells to grow out of control – they become **cancer** cells.

Ultraviolet radiation

OCR A	P2.1/2
OCR B	P1h
AQA	P1.13.5
EDEXCEL 360	P1b11

> **There is more about ozone on page 64.**

Ultraviolet radiation from the Sun travels towards the Earth. The **highest-energy** (highest-frequency) ultraviolet radiation is stopped by the **ozone layer**, a layer of the gas ozone in the upper atmosphere.

Unfortunately, using **CFC gases** in aerosol cans (as a propellant to force the contents of the can out of the nozzle) and as a refrigerant in refrigerators and freezers, caused CFC gas to rise in the atmosphere. This pollution **reacted** with the **ozone** and reduced the amount in the ozone layer.

Above the Antarctic an **'ozone hole'** developed and began to grow. International agreements have stopped the use of CFCs but the ozone layer will take time to recover. Where the ozone layer has been depleted, living organisms, especially animals, suffer more harmful effects from ultraviolet radiation.

The **lower-energy** ultraviolet radiation that passes through the ozone layer can cause **skin cancer** and damage the lens of the eye. The number of cases of skin cancer rose when people started to spend more time in the sun. **Light skins** are more **easily damaged** than **dark skins**. Dark skins absorb more of the ultraviolet radiation close to the surface; light skins allow the radiation to penetrate further into the body tissues.

Campaigns warn people to stay out of the sun during the hottest part of the day, and to cover up with a hat and clothes. They should also use a **high-protection factor sunscreen**, which reduces the ultraviolet radiation reaching the skin.

Sunscreens have a **sun protection factor (SPF)** number, which relates to how long a user can stay out in the sun. To find out the time you can stay in the sun wearing the sunscreen, **multiply the time** you could safely stay in the Sun **without** the sunscreen by the SPF. For example, if a person could safely stay in the sun for 10 minutes, using a sunscreen with SPF 15 means they can can safely stay in the sun for 150 minutes. It is important:
- not to miss any area of skin
- to put the cream on in a thick enough layer
- to reapply it according to the instructions.

Ulltraviolet blocking **sunglasses** are recommended to protect the **lens** of the **eye** from damage.

> **KEY POINT** Ionising radiation can cause cells to turn cancerous. Ultraviolet radiation from the Sun can cause skin cancer.

Using ultraviolet radiation

Ultraviolet radiation is used to detect **forged bank notes**. Bank notes have some features, for example the number of pounds, that **fluoresce** in ultraviolet radiation. This means they absorb the ultraviolet and re-emit visible light. You may have seen a detector in shops. The banknotes are placed under the light and the number shows up brightly, because it emits light if the bank note is genuine.

X-rays

OCR A	P2.1/2
OCR B	P1h
AQA	P1.13.5
EDEXCEL 360	P1b11

X-rays have **high energy** and can pass through the **body tissues**. They are **stopped** by **denser** materials such as the **bones** and by pieces of metal. They can be used to scan the body and show breaks in the bones. The diagram shows how a photographic plate can be placed behind the patient and X-rays directed towards the patient. The plate will darken where the X-rays strike it and leave white shadows of the bones, where the X-rays are absorbed.

The **operator** stands behind a **lead screen** so that he or she is not exposed to X-rays each time a patient is X-rayed. The patient should not receive too many X-rays. They are used when the **benefit** (for example, finding a broken bone) is greater than the **risk**.

In the 1950s, all pregnant women were X-rayed to check on the development and position of the baby. This caused a few cancers among children. The benefits were not greater than the risks and routine X-rays were stopped.

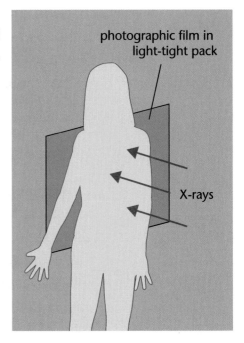

photographic film in light-tight pack

X-rays

Fig. 12.9 Taking an X-ray picture.

12.6 The atmosphere and global warming

Gases in the atmosphere

| OCR A | P2.3/4 |
| OCR B | P1h |

The Earth is surrounded by an atmosphere made up of different gases. The radiation from the Sun that can pass through the atmosphere is infrared radiation, light and ultraviolet radiation. Infrared and ultraviolet radiation are close to light in energy and frequency.

The radiation from the Sun provides energy for photosynthesis and warms the Earth's surface.

The Earth emits infrared radiation, but this radiation is at a low energy (low frequency). These energies are absorbed by some gases in the atmosphere such as **carbon dioxide**, **water vapour** and **methane**. This keeps the Earth warm. It is called the **greenhouse effect**. Without the greenhouse effect the Earth would be much colder – probably too cold for some species to survive.

In the last 200 years, the amount of carbon dioxide in the atmosphere has steadily increased. This is partly because we burn so much **fossil fuel**. We have also cleared many forests so that fewer trees are using the carbon dioxide for photosynthesis. This means that more of the infrared radiation is being absorbed and the Earth is warming up. This **global warming** causes **climate change**, which, scientists agree, is already taking place. There is less agreement about how much change is likely and what effect it will have. It could result in:

● extreme weather conditions in some regions
● rising sea levels due to melting ice and expansion of water in oceans, which may flood low-lying land
● some regions no longer able to grow food crops.

High-energy (high-frequency) infrared radiation from the Sun can pass through the glass into a greenhouse. Low-energy (low-frequency) infrared radiation from the plants cannot pass out through the glass. This keeps the greenhouse warm – the greenhouse effect.

Dust in the atmosphere

When a volcano erupts it produces a lot of gases and dust, which spread around the atmosphere. These reflect the radiation from the Sun and cause the Earth to cool.

If factories in cities produce large amounts of smoke and dust, these can reflect the radiation emitted from the city and keep it warmer.

HOW SCIENCE WORKS

Mobile phones

Before 1985, no one in the UK had a mobile phone, but by 2005, as shown in the table, 50 million people in the UK had, or regularly used, a mobile phone.

Year	Mobile phone users in the UK
1985	0
1990	Fewer than 1 000 000
1995	4 500 000
2000	25 000 000
2005	50 000 000

This means that scientists do not have enough data to say if using a mobile phone is harmful. They are studying sample populations to see if there are harmful effects, but it will be some time before they have conducted enough studies to be sure.

The **precautionary principle** says that we if we don't know about the effects of something we should not take the risk – 'Better safe than sorry.' In 2000, a group set up by the UK government called 'The Independent Expert Group on Mobile Phones' reported their findings on mobile phone safety. The following text summarises a Department of Health leaflet.

Risks

The radio waves that are received and sent by mobile phones transmit in all directions, to find the nearest base station. This means that some of the radio waves will be directed at the head of the person using the phone. These waves are absorbed into the body tissue as energy, which can eventually cause a very small rise in temperature in the head.

This effect is measured using specific absorption rates (SARs), which is a measure of the amount of energy absorbed by the body. The higher the SAR, the more energy your body is absorbing and the higher the rise in temperature.

Present research shows that the radio waves from mobile phones are sufficient to cause a rise in temperature of up to 0.1 °C. This does not pose a known risk to health. Some mobile phones have better SARs than others; you can find this information from your mobile phone manufacturer or retailer.

Children are thought to be at higher risk of health implications from the use of mobile phones. This is because their skulls and cells are still growing and tend to absorb radiation more easily.

Recommendations
You can minimise your exposure to radio waves:

- Only make short calls on your mobile phone.
- Children should use mobile phones only if absolutely necessary. Find out the relative SARs before you buy a new mobile phone.
- Keep your mobile phone away from your body when it is in standby mode.
- Only use your phone when the reception is strong. Weak reception causes the phone to use more energy to communicate with the base station.
- Use a mobile phone that has an external antenna. This keeps the radio waves as far away from your head as possible.

Sun and skin cancer

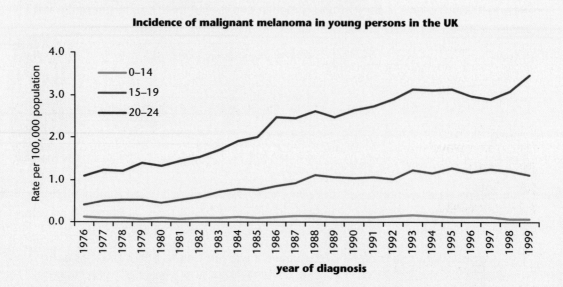

This graph shows that the number of cases of a type of skin cancer – malignant melanoma – has been steadily increasing. The graph may give the impression that 20–24-year-olds are at the most risk from the Sun. In fact this is because the damage to a child's skin from the Sun can cause cancer when they are older. In the UK we can learn from what has happened in Australia.

Australia is the country with the highest incidence of skin cancer in the world. Two out of three Australians will be treated for some form of skin cancer during their lifetime. This is partly because Australia has a generally light-skinned population with an outdoor lifestyle, and also it has clear skies, a depleted ozone layer and is close to the equator.

Data shows cases of melanoma began to rise in the 1930s. In the early 1980s, because the number of deaths from skin cancer was rapidly rising, the 'Slip! Slop! Slap!' programme encouraged fair-skinned Australians to protect themselves by slipping on a shirt, slopping on sunscreen and slapping on a hat. In the 1990s, Sun-protective clothing became popular because it was much easier to use than sunscreen. There was a 50% reduction in people getting sunburnt between 1988 and 1998.

The programme began to show success, in Australians under 60 years of age, when cases of melanoma fell from the mid-1980s to the mid-1990s. (Cases continued to increase in older Australians who had been exposed to the Sun before the 1980s.) Scientists hope that once the children born in the 1980s reach the age of 60, melanoma cases will have fallen to a very low level again. At the moment, after more than 60 years of steadily increasing deaths from melanoma, the trend has finally changed. This suggests younger Australians are now better at protecting themselves from the Sun.

Exam practice questions

1. Which of the following is ionising radiation?
    (a) infrared radiation
    (b) light
    (c) microwave radiation
    (d) ultraviolet radiation   **[1]**

2. Which radiation is best suited to communication between Earth-based stations and orbiting satellites?
    (a) microwave
    (b) radio
    (c) light
    (d) ultraviolet   **[1]**

3. How does light travel along optical fibres?
    (a) by diffraction
    (b) by dispersion
    (c) by refraction
    (d) by total internal reflection   **[1]**

4. Use some of the words below to complete these sentences:

    A signal can be transmitted using _____ radiation through an _____ _____ .

    An _____ signal picks up _____ which is amplified when the signal is amplified. A

    _____ signal is a code of 0s and 1s. Small added _____ signals do not have the value

    0 or 1 and can be set to 0 giving a cleaned up signal when the signal is _____.

    analogue   digital   infrared   light   noise   optical fibre   regenerated   **[7]**

5. Use some of the words below to complete these sentences:

    A microwave oven uses _____ which cause _____ molecules in the food to

    _____. This makes them heat up. Microwaves are _____ from the casing which is

    made of _____. The door has a metal grid in it to _____ the _____.

    Plastic and glass are _____ to microwaves so they can be used to contain the food.

    **absorbed   metal   microwaves   reflect   reflected
    transparent   vibrate   water**   **[8]**

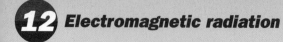

Exam practice questions

6. (a) Explain what is meant by 'wireless technology'.
    (b) What are two advantages of wireless technology?   **[3]**

7. Look at the How Science Works boxes about mobile phones and Sun and skin cancer.
    Scientists are setting up a study to see if brain tumours in children are increased by using mobile phones.

    Here are three possible samples for a 10 year study:

    A   100 children aged between 8 and 12: 50 use a mobile phone, 50 do not.
    B   1000 children aged between 0 and 15: 500 use a mobile phone at the start of the trial, 950 use one by the end.
    C   1000 children aged between 8 and 12: 400 use a mobile phone at the start, 600 use one at the end of the trial.

    (a) Give an advantage and disadvantage of sample A.   **[2]**
    (b) Give an advantage and disadvantage of sample B.   **[2]**
    (c) Explain how sample C could be used to produce a better sample of 800 for comparison.   **[2]**
    (d) Explain whether you think 10 years is a reasonable time for the study.   **[1]**
    (e) Some children use a mobile phone for several hours of conversation a day, others only use it to text. Explain how this will affect the trial.   **[2]**

The following topics are covered in this chapter:

- The Earth
- The Solar System
- Gravity and orbits
- Beyond the Solar System
- Observing and exploring
- The expanding Universe

13.1 The Earth

The structure of the Earth

OCR A **P1.1/2/4**

The **oldest** rocks on the **Earth** are about **4 thousand million years old**, which tells us that the Earth must be older than this. The rocks provide evidence for changes in the Earth.

Dating rocks

> You will find more details of the rock cycle in the chemistry section.

Rocks are **eroded** and the pieces are washed down into lakes and seas to form layers of **sediment**. These layers may contain dead animals and plants. The sediments are compressed into **sedimentary rocks** containing **fossils**. The newest rock is the top layer. Eventually all the continents would be worn down to sea level, but mountains are being continually formed. Over time, the rock layers may be lifted up and **folded** into **mountains**. They are then eroded again. Rocks can be dated by the order of the **layers**, the **fossils** they contain and by **radioactive dating**. (Because radioactive isotopes decay over time, the smaller the amount left in a rock, the older the rock.)

Plate tectonics

The Earth is made up of a **core**, a **mantle** and a **crust** as shown in the diagram.

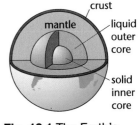

Fig. 13.1 The Earth's structure.

Once the world was mapped, people looked at the shape of the continents and thought they looked like pieces of a jigsaw puzzle. Alfred Wegener took this further, saying that fossils of plants and animals in America and Africa were similar, but modern plants and animals were not, suggesting that the continents had been joined in the past. He had no theory to account for how the continents moved or evidence that they were actually moving so the theory was not believed.

> **KEY POINT**
>
> **The theory of continental drift was first suggested by Alfred Wegener in 1915, but, owing to lack of evidence, it was not until 1967 that the theory of plate tectonics was accepted.**

The **Earth's crust** consists of a number of moving sections called **tectonic plates**.

- The **mantle** behaves like a very **thick liquid**.
- Heat from **radioactive decay** causes very slow-moving **convection currents** in the **mantle**.
- As the material from the mantle rises, it **melts** and becomes liquid **magma**.
- Magma flows out of the mid-ocean ridges, forming new rock.
- The **sea floor** spreads by about **10 cm a year** and this causes the continents to move apart.
- At other places where two **plates collide**, rocks are pushed up, forming new **mountain ranges**.

Evidence for this is found in the new rocks at the **ocean ridge**. The new rocks are rich in **iron**. Every few thousand years the Earth's **magnetic field** direction **reverses**. As the rocks solidify they are magnetised in the direction of the Earth's magnetic field. So the rocks contain a **magnetic record** of the Earth's field.

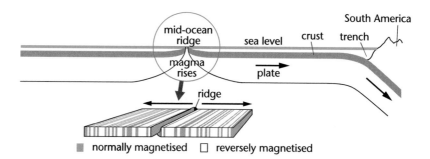

Fig. 13.2 The sea floor contains a magnetic record.

The Earth's magnetic field

OCR B   P2e

The **Earth** has a **magnetic field** as shown in the diagram. It can be thought of as if a large imaginary bar magnet lies inside the Earth with the south pole of the bar magnet at the Earth's magnetic North.

The north pole of a magnet is attracted to the south pole of another magnet. The name 'north pole' of a magnet means 'North of the Earth-finding end' of a magnet, in other words the North Pole of the Earth behaves as the south pole of a bar magnet.

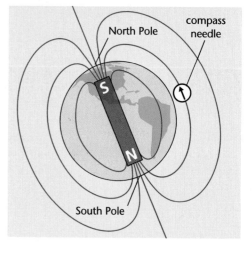

Fig. 13.3 The magnetic field of the Earth.

An **electric current** in a **coil of wire** generates a **magnetic field** as shown in Figure 13.4. **Plotting compasses** can be used to investigate the **direction** of a **magnetic field**. The compass needle is a small magnet that will line up with the magnetic field.

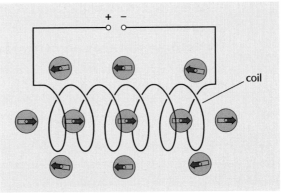

Fig 13.4 The magnetic field around a coil when a current passes.

> **KEY POINT**
>
> **The core of the Earth contains a lot of molten iron. As the Earth spins, electric currents flow in circles in the Earth's core and generate the magnetic field.**

Cosmic rays

OCR B   P2e

Cosmic rays are **ionising radiations** from space. They are fast-moving, charged particles. When they hit the atmosphere they can create **gamma rays**.

> **KEY POINT**
>
> **Moving charged particles (protons, electrons and ions) change direction when they move through a magnetic field.**

Cosmic rays change direction when they reach the **Earth's magnetic field**. They **spiral** round the **Earth's magnetic field lines**, and come down to Earth at the Poles. As they pass through the atmosphere they collide with gas molecules. As a result, the gas molecules emit coloured light. At the North Pole the lights are called the **aurora borealis**.

Solar flares

OCR B   P2e

> **KEY POINT**
>
> **A solar flare is a violent eruption in the Sun's atmosphere. Clouds of charged particles from the Sun are thrown out at high speed. They produce strong, changing magnetic fields. Some of these clouds of particles can travel towards the Earth.**

The fastest cloud of charged particles measured arrived at the Earth only 15 minutes after leaving the Sun. The **Earth's magnetic field** deflects many of the particles and **protects** us on the **Earth**, but the operation of artificial satellites in orbit around the Earth can be affected. If the **electronic circuits** in **artificial satellites** are **damaged**, satellites can break down, affecting telecommunications, navigation and weather prediction.

The magnetic field of the cloud of particles **disturbs** the **magnetic field** of the **Earth**, so that there are changing magnetic fields on the Earth. In the past, **electricity distribution networks** have **broken down** because of surges of electric current caused by the changing magnetic fields.

We now have satellites to give us some advance warning of clouds of particles heading towards Earth, and some equipment can be safely shut down.

The Moon

OCR B   P2e

The current theory of how our **Moon** was formed is that the **Earth collided** with another **planet**, about the size of Mars. Most of the heavier material of the other planet fell to Earth after the collision and the iron core of the Earth and the other planet merged. Some less dense material was thrown into orbit and formed the Moon. We have the following evidence for this:

● Samples of Moon rocks have been brought back to Earth by astronauts and the Moon **rocks** have the **same composition** of isotopes as Earth rocks – unlike rocks from other planets and moons.

● The Moon is completely made of **less dense** rocks – there is no iron core – unlike other planets, moons and asteroids.

● The Moon has **no recent volcanic activity** but its rocks are **igneous**.

13.2   The Solar System

The planets and other objects

OCR A   P1.1
OCR B   P2f
EDEXCEL 360   P1b12

The **Solar System** was formed over a very long time from **clouds** of **gases** and **dust** in space, about **five thousand million years ago**.

The **planets** orbit around the **Sun** (the star at the centre of the Solar System). Some of the planets are orbited by one or more **moons**. An object that orbits another is called a **satellite**. The Moon is a natural satellite of the Earth – and we have placed a number of **artificial satellites** in orbit around the Earth, and some around the Sun and other planets.

Asteroids are rocks that orbit the Sun. They vary in size – the largest is almost 1000 km across, some are 100 km across but many are as small as pebbles. They have been around since the formation of the Solar System. Most of these are between Mars and Jupiter. Jupiter is the largest planet and there is a large gravitational force towards it. This has prevented the formation of a planet between Mars and Jupiter, in the asteroid belt.

There are many **comets**. Some take less than a hundred years to orbit the Sun, others take millions of years. They are made of ice and dust. Most have a nucleus of less than about 10 km, which vaporises and becomes a cloud thousands of miles across when the comet is close to the Sun. Comets spend most of their time far from the Sun – much further away than the planet Pluto. **Meteors**, or shooting stars, are caused by dust and small rocks, usually from a **comet**. When the Earth passes through this debris in its orbit around the Sun, the dust and rocks are attracted by **gravity** towards the Earth. As they pass through the atmosphere, the pieces are heated and glow. Any pieces that land on the Earth are called **meteorites.**

Near Earth objects

OCR A — P1.1
OCR B — P2f
EDEXCEL 360 — P1b12

Near Earth objects (NEOs) are **comets** and **asteroids** that have been affected by the gravity of other planets so that they are now in an orbit that brings them close to the Earth. They are studied because their chemical composition is different from that of the Earth and they can give us information about the formation of the Solar System. Some of the NEOs could one day **collide** with **Earth**. The **craters** on the **Moon** are evidence of collisions in the past. **Craters** on the **Earth** have been mostly been **eroded** away, but the Barringer meteor crater can clearly be seen in the Arizona desert in the USA. A **collision** with a **large NEO** would result in a **crater** being formed and hot rocks being thrown up into the atmosphere. There would be widespread fires and the sunlight would be blocked by dust. This would cause **climate change** and many species would become **extinct**. The dust, containing unusual elements from the NEO, would settle as a layer onto the Earth's surface. The layer would be included in the new sedimentary rocks being formed, providing a record of the collision. There would be a smaller number of types of fossils above the layer (after the collision) than below the layer (before the collision). Evidence such as this points to the extinction of the dinosaurs being caused by a collision with a NEO.

> **KEY POINT** There is evidence that the extinction of the dinosaurs was caused by the collision of a large NEO with the Earth.

Monitoring NEOs

OCR B — P2g
EDEXCEL 360 — P1b12

Surveys by **telescope** observe and record the paths (trajectories) of all NEOs. They can be monitored from Earth or by satellite. We can make sure that we have advance warning of a collision. The idea of deflecting an NEO using explosions is being considered. At present, scientists are collecting information about the composition and structure of NEOs. The possibility of destroying one is only at the planning stage.

13.3 Gravity and orbits

Gravity

OCR A
OCR B   P2f/g
AQA
EDEXCEL 360   P1b12

KEY POINT The force of gravity is a force of attraction between two objects with mass. The force gets larger if the mass of either of the objects is increased. The force gets smaller as the distance between the two masses is increased.

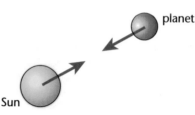

Fig. 13.5 The Sun attracts the planet with the same force as the planet attracts the Sun.

The **Sun** attracts the Earth and all the other **planets**. The **force of gravity** on a planet from the Sun keeps it moving in **orbit** around the **Sun**.

KEY POINT For an object to move in a circle, there must be a force on it towards the centre of the circle. This force is called the centripetal force.

KEY POINT If an object moves closer to the Sun, the gravitational force on it will increase and it will speed up.

Comets have very **elliptical orbits**. The force that keeps them in orbit is the gravitational force of attraction to the Sun but their distance from the Sun changes as shown in Figure 13.6. The **force on the comet** is **largest close** to the **Sun**, where the distance is smallest. The **speed** of the **comet** is much **larger when it is closer** to the **Sun**.

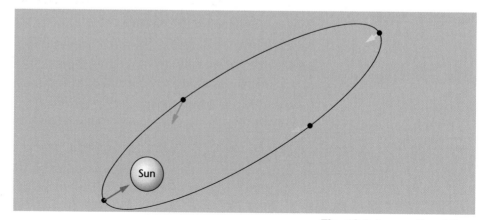

Fig. 13.6 A comet's orbit.

> **KEY POINT**
>
> This can be written as the equation:
>
> **Weight = mass × gravitational field strength**
>
> **The units of gravitational field strength are N/kg.**
>
> **Or**
>
> **Weight = mass × acceleration of free-fall**
>
> **The units of acceleration are m/s^2.**

On other planets, the gravitational field strength will be different – so **weight changes** but **mass** stays **the same**. In deep space where there is no gravity, objects will have zero weight.

13.4 Beyond the Solar System

Stars

OCR A P1.3
OCR B P2h
EDEXCEL 360 P1b12

Stars begin as large clouds of **dust**, **hydrogen** and **helium** – an **interstellar gas cloud**.

- The cloud is sometimes called a **nebula**. **Gravitational forces** between the particles make the nebula **contract**. This makes it heat up and it is now a **protostar**. As the core gets hotter, the atoms collide at high speed, losing their electrons.

- When the temperature is high enough, the **hydrogen nuclei fuse** together to form **helium nuclei**. This process is called **thermonuclear fusion**. When light nuclei fuse they release energy. **Light** and other **electromagnetic radiation** is also released. A **star** has been formed.

- The **star** is one of a large number of **main sequence** stars. The high pressure in the core is balanced by the attractive gravitational forces. Our **Sun** is a main sequence star. Stars spend a long time fusing hydrogen. Our Sun will do this for ten thousand million years. Eventually a star converts most of its hydrogen to helium. What happens next depends on the mass of the star.

- A **star of mass similar to that of** our Sun will eventually cool, becoming redder and expand to form a **red giant**. The core will contract and helium will fuse to form carbon and oxygen. After all the helium has fused, the star will contract and the outer layers will be gently lost. As these outer layers move away they look to us like a disc, which we call a **planetary nebula**. The remaining core becomes a small, dense, very hot **white dwarf**. This remnant core will then cool over an incredibly long time and eventually it will become a **black dwarf**.

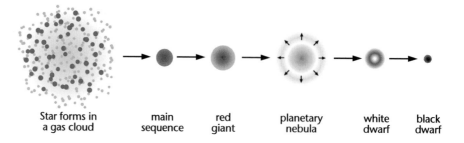

Fig. 13.7 The life cycle of a star of mass similar to that of our Sun.

> **All the elements heavier than helium were created in stars – everything on Earth – including us – is star dust.**

- A **more massive star** cools and expands to become a **red supergiant**. The core will contract and fusion in the core forms elements such as magnesium, silicon and iron. This fusion heats the star to a **blue supergiant**. When the nuclear fusion reactions are finished the star cools and contracts rapidly. The outer layers rebound violently against the dense core of the star. This is the **supernova**. It releases an enormous amount of energy. A supernova can be as bright as an entire galaxy. All the elements heavier than iron that exist naturally on planets were created in supernova explosions. The core is left as a **neutron star**. It has a large mass and is very dense.

- **Black holes** are the most dense neutron stars. They are so dense that even light cannot escape from their strong gravitational fields. This means we cannot see them. We can work out where they are because they attract gases from nearby stars. Matter accelerated towards the black hole gives out X-rays, which we can see.

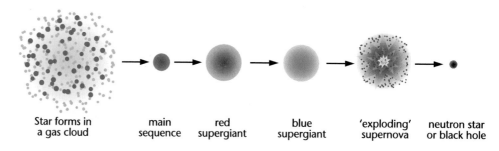

Fig. 13.8 The life cycle of a massive star.

Galaxies and the Universe

OCR A   P1.3
OCR B   P2h
AQA   P1.13.7
EDEXCEL 360   P1b12

Our Sun is a star in the **Milky Way galaxy**.
Galaxies are:

- collections of thousands of millions of stars
- all moving away from us, and from each other.

The Milky Way galaxy is shaped like a flat disc. There are thousands of millions of galaxies in the **Universe**. Between them there are empty regions that would take light hundreds of millions of years to cross.

The Universe is everything that exists: there is nothing outside the Universe – not even empty space.

13.5 Observing and exploring

Size: distance and time

The light year

Light travels through a vacuum at 300 000 km/s. Light from the Sun spreads out on a journey in all directions. The small amount we see on Earth arrives after about 10 minutes. After a **year**, **light** from the Sun will have **travelled** a distance of 300 000 km/s times the number of seconds in a year.

> **KEY POINT**
> A light year is the distance light travels in a year. Distances to stars and galaxies are so large that we measure them in light years. The nearest star to Earth is the Sun. The second nearest is about 4 light years from Earth.

A light year is about 9.5 × 10¹² km, in other words nine and a half million million kilometres. You don't need to remember this number, but you need to understand how it is worked out.

When you look through a telescope and see a star 100 light years away, what you see – the light entering your telescope – **left** the star **100 years ago**. So looking at very distant planets is like **looking back** in **time**. **Distant** objects look **younger** than they really are.

If an alien 950 light years away looks at Earth through its telescope on the right day in 2016 it will be able to watch the Battle of Hastings in 1066.

Object	Details of size	Details of age
asteroid	wide range, from pebble to 1000 km across	
the Moon	smaller than Earth (diameter about a quarter of the Earth's)	4500 million years
smallest planet Pluto	smaller than the Moon	
Earth	diameter 12 760 km	5000 million years
largest planet Jupiter	diameter over 10 times the Earth's	4600 million years
the Sun	diameter over 100 times the Earth's	
Solar System	diameter about ten thousand million km (10 000 000 000 km) (light would take between 10 and 11 hours to cross the Solar System)	
nearest star	about four light years	
Milky Way galaxy	diameter about 100 000 light years	
the observable Universe	about 14 thousand million light years	About 14 thousand million years

Difficult observations

OCR A   P1.3

All the **information** we have about objects outside the Solar system comes from **observations** made with **telescopes**. What we know depends on the **electromagnetic radiation** from the stars and galaxies.

It is very difficult to work out how far away a star or galaxy is. A very bright star may look bright because:

- it is larger than other stars
- it is hotter than other stars
- it is closer than other stars.

If the colour of two stars is the same and there are reasons to believe they are similar stars, then a difference in brightness can be used to measure the distance. The **further away** the star, the **dimmer** it is.

Parallax

You may have noticed this effect from a train or car window; objects close to you seem to move more quickly, and change position, when compared to distant objects.

Another method is to use **parallax measurement**. The Earth orbits the Sun. If an **observation** of the night sky is made, and then we wait for **six months**, the Earth will have **changed** its **position** in space by a distance equal to the **diameter** of its **orbit**.

When the night sky is observed from this position, a **star** that is **close to Earth** will have **changed its position** when **compared** with more **distant stars** in the **background**. The amount the near star has moved can be used to calculate how far it is from the Earth. This is shown in Figure 13.9.

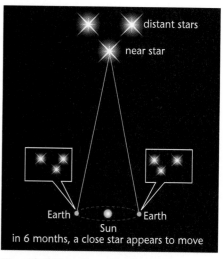

Fig. 13.9 How to calculate the distance a star is from the Earth.

These difficulties in making observations lead to **uncertainty** in our **measurements** of the distances to stars and galaxies.

Light pollution

Another difficulty for astronomers is the amount of **light pollution** from the Earth. All the light from our cities shines into the night sky, which makes it difficult to pick up the very **weak signals** from **distant stars** and **galaxies**. To see the Milky Way (it looks like a milky strip of stars in the sky), you need to be far away from towns and cities.

Space travel in the Solar System

OCR B   P2f
EDEXCEL 360   P1b12

Manned missions

We have used **spacecraft** to put men and women in **orbit** and land men on the **Moon** and bring them home. There are plans to send humans to **Mars** but the difficulties of space missions to other planets are extreme:

- Enough **food, water, oxygen** and **fuel** must be carried for the entire trip, including any delays. The distances involved mean that emergency supplies could not be sent in time to be of use. An **artificial atmosphere** must be set up inside the spacecraft, and levels of carbon dioxide and oxygen have to be monitored.
- Interplanetary space is very cold. There must be **heating** to keep astronauts warm. As the spacecraft will receive direct radiation from the Sun, it will heat up, so **cooling** is required.
- There will be very **low gravity** during the journey. This will affect health. Bones lose density and muscles waste away under these conditions. A special exercise programme will be needed using **exercise machines**. Experiments have been done in creating artificial gravity by spinning spacecraft, but have not been very successful.
- Some way of **shielding** astronauts from **cosmic rays** is needed. The radiation released when there is a solar flare is deadly (see p. 203).
- **Distances** are so large that only the closest planets could be reached.

Unmanned missions

Unmanned spacecraft have some advantages. They can operate unharmed in a lot of conditions that would kill humans. By using **remote sensors** and computers they can send back information on:

- temperature
- magnetic field
- radiation
- gravity
- gases in the atmosphere
- composition of the rocks
- appearance of the surroundings (using TV cameras).

The **costs** of unmanned missions are **less** and there are **no lives at risk** if something goes wrong. The NASA 'Viking' spacecraft and the 'Spirit' and 'Opportunity' Mars rovers have all studied the rocks and terrain of Mars. We have a lot of information about Mars without any humans travelling there. The disadvantages are that everything must be thought of before the mission leaves – **no adjustments, repairs** or changes can be made unless they are programmed into the computers and can be done remotely. Most people are not as interested in unmanned missions, and they do not inspire people in the way that manned missions do.

Using telescopes

AQA   P1.13.7

Beyond the Solar System **telescopes** are the only way to study the Universe. Even in the Solar system we have obtained a lot of information from telescopes. Telescopes can be positioned **on Earth** where they are **easier** to **maintain** and repair, but where people must look **through** the **atmosphere**. Some telescopes are positioned **high up** on mountain tops to get **above** the clouds, dust and pollution, and away from city lights. Telescopes in **orbit** avoid these problems and have a much clearer view of the stars. **Launching** a telescope into orbit is **expensive**.

There are **different telescopes** to use **all** the different **ranges** of the **electromagnetic spectrum**, including **radio telescopes**, which observe radio waves from space. As X-rays do not pass through the atmosphere, **X-ray telescopes** are always placed in orbit.

Life elsewhere in the Universe

Astronomers have identified some distant **stars** that have **planets** around them. These planets could be home to **alien life**. We have **not** yet discovered any trace of **alien life**, either existing now or that lived in the past. Even if only a few stars have planets, because there are so many stars, scientists think it **likely** that life has **evolved somewhere else** in the Universe. A life-form like ours would need a **planet close** enough to its **Sun** to be warm but not so close it would burn. The alien Sun would need to be in a stable part of its life – which probably means a **main sequence star**.

Even if we discovered life elsewhere it is difficult to see how we could **make contact** with the large distances and **time delays** for signals – any signal we receive from a million light years away was sent a million years ago.
The **Search for Extra-Terrestrial Intelligence** (SETI) is a project that scans **radio waves** from space for evidence of **patterns** that suggest aliens using radio waves for communication – as we do.

13.6 The expanding Universe

The Big Bang

The **Universe** began with a 'Big Bang' about **14 thousand million years ago**. It is still **expanding**. It is difficult to predict whether it will continue to expand, stop, or start to contract.

For Edexcel you need to know about the alternative theory to the Big Bang described in the How Science Works section on page 214.

KEY POINT

Hubble's Law:

The recessional speed of a galaxy is directly proportional to its distance from our galaxy. The further away a galaxy is, the faster it is moving away from our galaxy.

Hubble's Law is strong evidence for the Big Bang because it suggests that the Universe started expanding from a small starting point.

Microwave background radiation

OCR A   P1.3
OCR B   P2h
AQA   P1.13.7
EDEXCEL 360   P1b12

Do not confuse this with radioactive background.

Microwave background radiation comes from all parts of the **Universe** and is the left over radiation from the **Big Bang**. It has **cooled** to the **microwave** region of the electromagnetic spectrum.

All the information we have about distant stars and galaxies comes from the radiation they produce.

Red shift

OCR B   P2h
AQA   P1.13.7
EDEXCEL 360   P1.12

If a **source of waves** is moving **away**, the **wavelength** appears **longer** and the **frequency** of the waves appears **lower**. Fig. 13.10 shows how in the time between one crest and the next the source moves further away so the wavelength is longer than for the stationary source.

Fig. 13.10 Waves from a moving source.

If a source of light waves is moving away so that the wavelength is **longer**, this is called a **red shift**.

A red shift in the light from a star shows that the distance between us and the star is increasing. The **bigger** the **red shift**, the **faster** the star is **moving away**.

Scientists think that the red shift we see for all galaxies is because space is expanding, not because we, or the star, are moving through space.

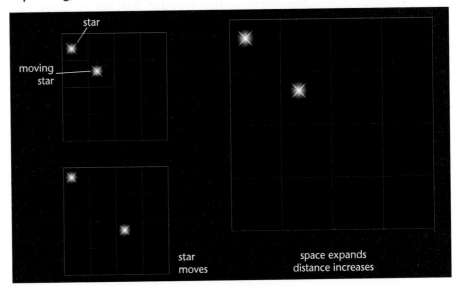

Fig. 13.11 The difference between a moving star and expanding space.

HOW SCIENCE WORKS

Ideas about the Universe

In 1920, there was a 'great debate' between two astronomers. Each gave a talk about ideas that some astronomers had at the time.

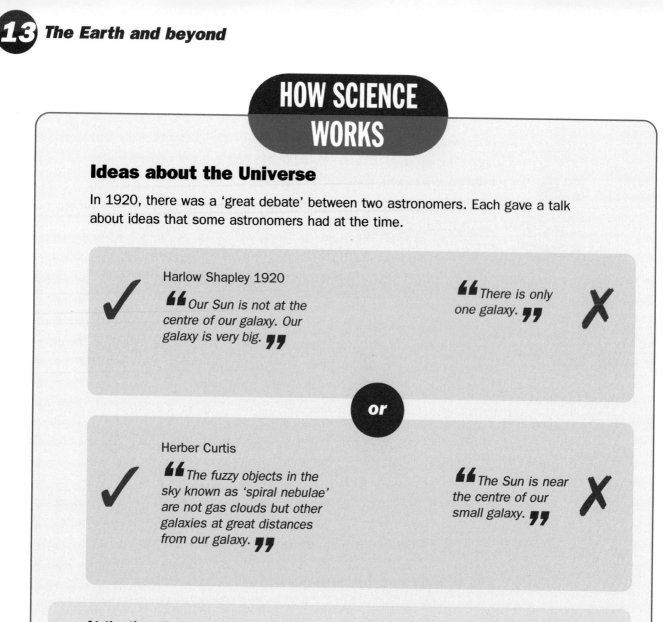

Harlow Shapley 1920

✓ *"Our Sun is not at the centre of our galaxy. Our galaxy is very big."*

"There is only one galaxy." ✗

or

Herber Curtis

✓ *"The fuzzy objects in the sky known as 'spiral nebulae' are not gas clouds but other galaxies at great distances from our galaxy."*

"The Sun is near the centre of our small galaxy." ✗

At the time there was no evidence to prove who was right or who was wrong.

Evidence: in 1924, Edwin Hubble showed that, as Curtis thought, the distance to 'spiral nebulae' was much greater than the size of the Milky Way. In the 1920s and 1930s, the size of our galaxy was measured more accurately, showing, as Shapley thought, it was large and the Sun was not central.

Edwin Hubble also used the red shift to show that all galaxies are moving away from all other galaxies. Those furthest away were moving fastest.

New ideas

George Gamow 1946

"If all galaxies are moving away from each other, then at some time in the past they must all have started from the same point. This 'explosion' was the beginning of the Universe. The 'explosion' would produce radiation. By now it will have cooled to microwaves."

HOW SCIENCE WORKS

Fred Hoyle 1950

" Steady state theory makes more sense. The Universe must always have existed. I prefer the steady state theory to an 'explosion': it's just a 'Big Bang' theory. "

Evidence: Astronomers started to look for the microwave radiation. In 1965, Arno Penzias and Robert Wilson discovered that a low level of microwave radiation came from all directions in the Universe. They were not looking for the radiation and did not know what it was. They had found the cosmic microwave background which supports the Big Bang theory.

The next new ideas

" The expansion of the Universe will slow down, and it will contract to a point, because of the gravitational attraction between all the mass in the Universe. There is enough mass in the Universe for this to happen. "

" After the Universe has contracted to a point there will be a new Big Bang. This is called the oscillating Universe theory. "

or

" The Universe will continue to expand forever, because there is not enough mass to provide the gravity to reverse the expansion. "

or

" There is just enough mass to stop the expansion of the Universe, but not to start it contracting again. "

The evidence scientists need is how much mass there is in the Universe. They know there is some mass they cannot see. This is because objects they can see are reacting to larger gravitational forces caused by the mass they cannot see. Scientists call this Dark Matter.

What is Dark Matter   How much is there ?

Scientists are searching for evidence to support the different ideas.

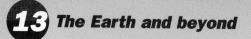

Exam practice questions

1. Put these objects in order of size, starting with the smallest.

   Universe   Sun   Milky Way galaxy   Solar System   Earth   **[1]**

2. Which of these best describes the way galaxies move?
   - **(a)** Galaxies are stationary, but our galaxy is moving away from them.
   - **(b)** Galaxies are all moving away from each other at constant speed.
   - **(c)** Galaxies are all moving away from each other. Those furthest away are moving fastest.
   - **(d)** Some galaxies are moving away from our galaxy (the Milky Way).   **[1]**

3. Complete this description of the life of a star like our Sun:

   A large cloud of dust and gases called a _____, or an interstellar gas cloud, _____ and heats up, forming a _____. In a new star, _____ nuclei join in a process called _____ _____ to form _____ nuclei. The star will be a _____ _____ star for a long time.

   When all the _____ has been fused, the star becomes a _____ _____. As it cools, the outer layers move away as a _____ _____, leaving a _____ _____ which will eventually cool to a _____ _____   **[12]**

   black   contracts   dwarf   fusion   giant   helium   hydrogen   main   nebula
   nuclear   planetary   protostar   red   sequence   white

4. The current theory of how the Moon was formed suggests that a planet collided with the Earth and the less dense rocks were thrown up as the Moon. Give two pieces of evidence that support this theory.   **[2]**

5. When we observe light from a distant star we see a **red shift** in the wavelength.
   - **(a)** What is meant by red shift?
   - **(b)** What causes a red shift?   **[2]**

6. Look back at the 'Ideas about science' section.
   - **(a)** What two pieces of evidence are there for the Universe starting with a Big Bang?
   - **(b)** What ideas do scientists have about what will happen to the Universe in the future?
   - **(c)** What will decide which of these ideas may occur?
   - **(d)** Explain whether it is possible for scientists to find the answer to what happened before the Big Bang.   **[6]**

Answers

Chapter 1

1. **(d)**
2. **(a)**
3. **(a)**
4. Growth; amino acids; iron; digested; absorption
5. **(a) (i)** Pancreas   **(ii)** liver
   (b) Needed for respiration by cells; if it becomes too high it passes out in the urine
   (c) Five from: glucose diffuses into the blood stream/in the small intestine/rise in blood sugar level causes release of insulin/from the pancreas/causes liver to store more glucose/as glycogen
6. **(a)** Hormone A: oestrogen; first half
   Hormone B: progesterone; second half
   (b) Ovaries
   (c) Higher in the second half of the cycle; small fluctuations
   (d) (i) Largest increase in temperature around time of ovulation; intercourse then is more likely to lead to pregnancy
      (ii) Avoid intercourse around the time of the temperature increase
   (e) Women do not all ovulate regularly/may have intercourse the day before a temperature increase/does not give any protection against STDs
7. **(a)** Two from: it is unlikely that a large number of embryos will survive if left to develop/removing some may allow the others to survive/this would result in the destruction of some of the embryos/some people would consider this immoral
   (b) Two from: less fit to give birth/less fit to bring up children/more likely to die before the children have grown up/may find it harder to relate to the children as they grow up
   (c) Chance of success after one treatment is small/chance increases with number of treatments; but only one treatment would allow others a chance

Chapter 2

1. **(b)**
2. **(a)**
3. **(d)**
4. Pathogen; lysozyme; acid; toxins

5. **(a)** Measles can cause babies to die/mumps can cause deafness in young children
   (b) Because if they catch rubella their babies may become brain damaged
   (c) Three from: contains a weakened or dead pathogen/stimulates the production of antibodies/from white blood cells/memory cells are made/if the live pathogen invades it can be killed rapidly
   (d) It often contains a live but weakened pathogen; people get a mild form of the disease
6. **(a)** Nicotine
   (b) Cannot do without it; lack of the drug causes withdrawal symptoms
   (c) Slow it down
   (d) Different penalties for illegal possession; because some are more dangerous than others
7. **(a)** To see if they work; to see if they are safe
   (b) So that a person does not know if they have taken a drug or not; to eliminate psychological effects
   (c) For: animals were suffering unnecessarily; they do not always react in the same way as people
   Against: this is only one case whereas thousands of drugs have been tested; testing on animals may have saved thousands of lives

Chapter 3

1. **(c)**
2. **D**
3. **(a)**
4. Nucleus; chromosomes; DNA; proteins
5. **(a)** There would only be 23; they would not be in pairs
   (b) The last pair are different; contains an X and a Y chromosome
6. **(a)** Recessive; Jackie and Leroy have the allele but not the disorder
   (b) Leroy's gametes are F and f; correct offspring: FF, Ff, fF, ff
   (c) 25%/1 in 4/a quarter
   (d) Would know for certain if she was expecting a baby with cystic fibrosis; she would then have to decide whether to have a abortion or not

7. **(a)** Contradicted the Bible; he knew most
people were very religious; worried that
they would be very upset
   (b) Some species must have died out; these
were likely to be the less well adapted;
could find some fossils that were similar to
organisms living today
   (c) A theory is an idea or an explanation; data
is factual information; a theory has not
been proved correct

Chapter 4

1. **(c)**
2. **(b)**
3. **(d)**
4. Kingdoms; vertebrates; mammal; species
5. **(a)** Two from: store of food or energy/do not
need to eat so often/helps insulate the
body from the sun
   (b) Deep roots can absorb water from deep
underground; roots that are spread out
can absorb water over a wide area; shallow
roots can absorb the water before it
evaporates
   (c) Large animals cool down more slowly;
small ears reduce the surface area
   (d) Less competition between the larvae and
adults
6. **(a)** Three from: place quadrat at
random/count the number of animals in
quadrat/repeat and take an
average/multiply the number of organisms
by the number of quadrats that would
cover the whole field
   (b) (i) Three from: ladybirds eat greenfly/
greenfly increased in July because
plenty of food and warmth/more food
for ladybirds so their numbers
increased/greenfly numbers dropped
because more were eaten so ladybird
numbers started to drop
   (ii) predator–prey graph
7. **(a)** Spread over large areas; difficult to track
them/find them all
   (b) Two from: only hunt for scientific
research/vary the type of whales being
hunted/but have caught fin whales that are
protected/some people are not convinced
that it is for scientific research
   (c) It would keep their culture alive; they
would not kill many whales

Chapter 5

1. **(c)**
2. **(b)**
3. **(d)**
4. freezes, 100°C, compound, hydrogen, solvent,
isotope, protons, neutrons, hydroxide,
potassium
5. **(a)** The smallest part of an element that can
exist
   (b) A neutral group of atoms held together by
covalent bonds
   (c) The elements of a vertical column in the
periodic table
6. **(a)** Copper
   (b) Air
   (c) Brass
   (d) Sodium chloride and water
   (e) Calcium carbonate, sugar and water
   (f) Calcium carbonate and sugar
7. **(a)** Physical: **(i)**, **(ii)** and **(iii)**
Chemical: **(iv)** and **(v)**
   (b) Metal – it conducts electricity and reacts
with water to give an alkaline solution
   (c) In oil
   (d) (i) Group 1
   (ii) Any two from lithium, sodium,
potassium, rubidium, caesium and
francium
   (e) (i) White
   (ii) Le^+ and Cl^-

Chapter 6

1. **(d)**
2. **(a)**
3. **(a)**
4. mixture, hydrogen, alkanes, alkenes, bromine,
colourless, polymers, monomers, cracked,
petrol
5. **(a)** Nitrogen
   (b) Carbon monoxide
   (c) Methane
   (d) Sulphur dioxide or oxides of nitrogen
   (e) Carbon dioxide, carbon monoxide, oxides
of nitrogen
   (f) Oxides of nitrogen
   (g) Carbon dioxide and methane
6. **(a)** potassium nitrate KNO_3, ammonium
nitrate NH_4NO_3, urea $CO(NH_2)_2$
   (b) potassium phosphate K_3PO_4
   (c) (i) Excess fertiliser dissolves in rain water
and drains into rivers. Algae grow well
on the fertiliser and cover the river.
The algae die and bacteria
decompose them. The bacteria use up
most of the oxygen in the water.
There is little oxygen left for the fish
and they die.
   (ii) eutrophication.

7. **(a) (i)** $N_2 + 3H_2 \rightarrow 2NH_3$
      (ii) Iron
   (b) Reaction does not go to completion or reaction is reversible.
   (c) Ammonia has the highest boiling point and it turns into a liquid while N_2 and H_2 remain as gases.
   (d) (i) 350°C and 400 atmospheres
      (ii) Reaction too slow at this temperature and expensive to build factory to withstand 400 atmospheres
      (iii) 450°C and 200 atmospheres

8. **(a) (i)** A compound of carbon and hydrogen only
      (ii) Alkane – it fits the general formula C_nH_{2n+2}
   (iii) Add bromine water – the solution changes from reddish brown to colourless
   (b) Nitrogen
   (c) (i) Complete combustion
      (ii) Incomplete combustion
      (iii) Nitrogen reacting with oxygen
      (iv) Unburnt fuel
   (d) $C_8H_{18} \rightarrow C_8H_{16} + H_2$
   (e) (i) $N_2O + CO \rightarrow N_2 + CO_2$
      (ii) Carbon monoxide is very poisonous

9. **(a)**

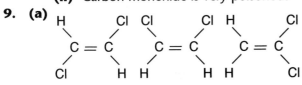

   (b)

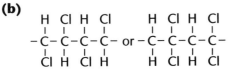

   (c) Addition polymerisation
   (d) Carbon dioxide and hydrogen chloride

Chapter 7

1. **(b)**
2. **(b)**
3. **(b)**
4. metal, useful, separate, solder, lower, electrical, soft, ductile, malleable, harder
5. **(a)** Sulphuric acid
   (b) Phosphoric acid
   (c) Potassium hydroxide
   (d) Ammonia solution
6. **(a) (i)** Continental
      (ii) Oceanic trench
      (iii) Mantle
   (b) (i) Oxygen
      (ii) Iron
   (c) (i) Denser
      (ii) Subduction
      (iii) Earthquakes, tsunamis, volcanoes

7. **(a)** A because it can be manufactured cheaply since it is abundant and easily extracted (probably iron)
   (b) B because it is rare and unreactive; it is economically worth recycling
8. **(a)** Fe
   (b) The least reactive
   (c) (i) Conduct electricity
      (ii) Hydrogen
   (d) (i) Cost of manufacture of aluminium is much higher
      (ii) It is very rare
   (e) Either unreactive elements and therefore occur in the free state or easily reduced to elements
   (f) Aluminium chloride + potassium → aluminium + potassium chloride
      $AlCl_3 + 3K \rightarrow Al + 3KCl$
   (g) (i) 'Cannot be renewed' or 'eventually they will be used up'
      (ii) The recovery and processing of materials after they have been used
      (iii) Economic advantage: savings of energy/ better conservation of natural resources; environmental advantage: more effective waste disposal/helps to solve the problem of litter accumulation
      (iv) Costly/difficulty of separating materials
9. **(a)** electrical conductivity
   (b) (i) does not break
      (ii) not easily scratched/unaffected by sunlight/cheaper
   (c) (i) does not rust, less dense
      (ii) plasticiser
   (d) less reactive
   (e) (i) good conductor of heat/does not burn/does not melt
      (ii) poor conductor of heat/catch fire/melt
   (f) carbon monoxide and hydrogen chloride
   (g) polystyrene will be linear; expanded polystyrene will have cross linkages.

Chapter 8

1. **(c)**
2. **(b)**
3. **(c)**
4. marble, loss, volume, time, salt, temperature, smaller, the same, faster, larger
5. **(a)** Endothermic
   (b) Exothermic
   (c) Bond-making – the reaction is exothermic, therefore more energy given out when bonds form than taken in when bonds break.
6. **(a)** Less heat loss; alcohol burns completely
   (b) Mass of spirit lamp at start/mass of spirit lamp at finish/mass of water/temperature of water at start/temperature of water at finish

(c) Copper is a good conductor of heat; polythene is a very poor conductor of heat

(d) To prevent alcohol evaporating when experiment was finished

(e) Energy given out when bonds formed is greater that the energy required to break the bonds

(f) (i) C_4H_9OH

(ii) 38.0 (kJ/g)

(g) Cost of the alcohol

7. (a)

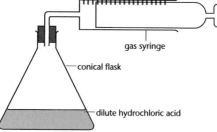

(b) (i)

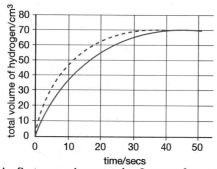

(ii) in first experiment take 9 away from all the results

average the two results

but ignore volume at 20 seconds in experiment 2

(c) at the start – slope of tangent to graph is greatest

(d) 70 seconds

(e) 15 seconds

(f) see graph in (b) (i) – steeper slope; volume of gas given off the same

Chapter 9

1. **(d)**
2. Nuclei, unstable
3. As low as reasonably achievable
4. **(c)**
5. **(c)**
6. Fuel, rods, nucleus, neutron, neutrons, fission, chain, control, rods, neutrons, coolant
7. High, low
8. **(a)** Beta radiation

   (b) No alpha radiation would get through the paper. All the gamma radiation would get through the paper. Only beta radiation intensity would vary with thickness

9. A radioactive dose that emits gamma radiation is given to the patient. The source builds up in the tumour. The gamma radiation kills the cancer cells. Alternatively, a beam from a gamma source is directed at the tumour

10. **(a)** One-sixteenth

    (b) (i) It would decay away before reaching the organ and being recorded by the gamma camera

    (ii) The patient would have radioactive material in him for days

11. **(a)** A nucleus splits in two

    (b) A million times more in a nuclear reaction

12. The precautionery principle means that you should not take a risk if you are not sure of the consequences. It can be summarised as 'Better safe than sorry'.

Chapter 10

1. **(c)**
2. **(c)**
3. **(b)**
4. **(a)**
5. 8400, 0.5, more than
6. Conduction, convection, radiation (any order), convection, currents, cavity, walls, air, gaps, radiation
7. Coal, gas, nuclear, carbon dioxide, fossil fuel, wind energy, hundreds, nuclear, nuclear (or radioactive) waste, radioactive, millions
8. Alternating, frequency, fifty
9. Energy, unit, unit, 60p
10. **(a)** Carpet is an insulator (or stone is a better conductor)

    (b) The layer of air between is a good insulator

    (c) Convection currents carry smoke upwards

    (d) Hot water rises by convection and the cold will fall and be heated

    (e) To lose heat from the pipes at the back

11. **(a) (i)** Air stops conduction

    (ii) Stops conduction and convection

    (iii) Stops radiation

    (b) Stops heat from being transferred to the food, as well as heat being transferred from the food

12. **(a)** 58%

    (b) Burns gas not coal, uses hot gas and steam, coal uses steam only, *or* more efficient (58% compared to about 40%)

    (c) Produces carbon dioxide which causes global warming *or* produces pollution, e.g. sulphur dioxide *or* is a fossil fuel so will run out

13. **(a)** The amount of wind (or wind speed)

    (b) C

    (c) (i) The appliances used at that time of day – e.g. cooker

(ii) At night everything might be switched off

(d) If the wind drops below the maximum the turbine will produce less.
Will not cope with peaks in demand

(e) Can use when there is no wind or to give a more constant supply

(f) (i) 1500 **(ii)** 500

(g) The time for the turbine to save the amount of money that it cost to buy and install

14. **(a)** 1748 units
(b) £174.80
(c) 6
(d) 60p
(e) the lamp
(f) 5p

Chapter 11

1. **(b)**
2. **(b)**
3. **(a)**
4. disturbance, medium, energy, matter, vibrates
5. speed, wavelength, frequency, direction
6.

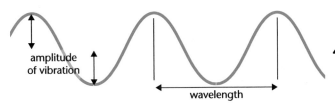

7. **(a)** 10
(b) $v = \lambda f$
(c) 2000 m/s or 2 km/s

Chapter 12

1. **(d)**
2. **(a)**
3. **(d)**
4. infrared, optical fibre, analogue, noise, digital, noise, regenerated
5. microwaves, water, vibrate, reflected, metal, reflect, microwaves, transparent.
6. **(a)** Communication using electromagnetic waves without wires
(b) Two of: We can receive phone calls and e-mail 24 hours a day; no wiring is needed to connect laptops to the Internet, for mobile phones or radio communication with wireless technology is portable and convenient

7. **(a)** Same numbers at the start and end for each group (phone users and non-users). Very small sample – results could be just random chance
(b) Bigger sample will give more meaningful results – but by the end of the trial sample not using phone is small; 0–2-year-olds cannot use phones
(c) If the 200 who start to use a phone are removed there is a sample of 800, 400 who use a phone and 400 who don't
(d) Not really – as shown by the Sun cancer case it could take longer for effects to be seen
(e) Texting does not expose the brain to so many microwaves, so could make using a mobile phone look safer than it is. A good study would need to know more about the actual phone use

Chapter 13

1. Earth, Sun, Solar System, Milky Way galaxy, Universe
2. **(c)**
3. nebula, contracts, protostar, hydrogen, nuclear, fusion, helium, main, sequence, hydrogen, red, giant, planetary, nebula, white, dwarf, black, dwarf
4. Two of: (i) Moon rocks have the same composition as Earth rocks; (ii) the Moon is made of less dense rocks – there is no iron core – unlike other planets, moons, and asteroids; (iii) the Moon has no recent volcanic activity but its rocks are igneous
5. **(a)** The wavelength of radiation (e.g. light) appears longer than it is
(b) Movement of the star away from the observer
6. **(a)** Microwave background radiation and galaxies moving away from each other
(b) It will expand for ever, or stop expanding, or start contracting. It may oscillate – expanding and contracting again
(c) The amount of matter in the Universe
(d) No: there is no evidence of what happened before

Index

absorption 10
accommodation of eye 13
acid rain 64, 161
acids 64, 101, 161
active packaging 88
adaptations 58–9
addiction 32
aerobic respiration 11
Agenda 21 (Earth Summit) 67
air 102–3
alcohol (drinking) 32–3
alcohols 100–1
alkali metals 83–4
alkanes 95–6, 97
alkenes 96–7
alleles 43
alloys 121–3
alpha (α) emission/radiation 145, 147, 150
alternating current (a.c.) 166
aluminium 118–19, 140–1
amalgam 122
ammonia 104–5, 132
anaemia 36
anaerobic respiration 11
analogue signals 188–9
angina 35
animal kingdoms 56
anions 73
anorexia 36
antibiotics 29–30
antioxidants 87
antiseptics 30
asexual reproduction 42
asteroids 204
atmosphere 102, 195–6
atoms 72–3, 78

background radiation 145–6
Bacteria (species) 56
Bakelite 99
balanced diet 9
bank notes 194
batch process 100
bauxite (aluminium ore) 118, 141
HMS Beagle 49
beta (β) emission/radiation 145, 147–8, 150
big bang theory 212
binomial system 57
biodiversity 66
biomass 163
blood pressure 34–5
blood sugar control 18
body mass index (BMI) 34
body's defences 28
boiling points 157
bond making and breaking 136–8
bonding 74–6
brain 16

brass 122
bread 77
bronchitis 33
burning 131
butane 98

cacti 59
calcium cycle 93–4
calorimetry 138
camels 59
cancer 47
car components 108
carat (definition) 122
carbon 91, 92, 95–8
carbon dioxide 91–2, 132
carbon monoxide 33
carboxylic acids 101
catalysts 134, 135
catalytic converters 103
cations 73
cavity wall insulation 159
cement 93–4
central nervous system (CNS) 14–15
centripetal force 206–7
cerebral hemisphere 17
CFCs 64, 194
chain length, polymers 115
chain reactions 150
chemical properties
    group 1 elements 84
    group 7 elements 85
chemical reactions 131
chemical transmitters (neurones) 15
chlorine 75, 84, 132
chromatography 80–1
chromosomes 41–4
classification, organisms 55–6
climate change 196
cloning 45–6
CNS see central nervous system
co-ordination 12
cold 19
combustion 131
comets 204–5, 206
communications 8, 188–9, 192–3
communities 57
competition 58–9
compounds 75–6, 77–8
concentration 134–5
concrete 93–4, 115
conduction 158
conservation 66–7, 69
continental plates 113
contraceptives (oral) 22
convection 113, 158–9
cooking 79–80
copper 121
core of earth 112
cosmic rays 203–4
covalent bonds 75–6, 95–6, 116

cracking 98, 131
creation of life 48, 51
critical angle 182
cross-linking of polymers 115
crude oil 94–5
crust of earth 112
crystallinity in polymers 115
Curtis, Herbert 215
cystic fibrosis 44

dangers of radiation 146–7
Dark Matter 216
Darwin, Charles 49, 51
dating rocks 201
defences of the body 28
deficiency diseases 36
deforestation 65
denaturation 80
development (sustainable) 67
diabetes 18
diamond 139
diet 9
diffraction of waves 181
digestion 9–10
digital signals 188–9
dinosaurs 48
direct current (d.c.) 166
discoveries 107
displacement reactions 86, 131
distances (eyes) 14
distillation (fractional) 94–5
DNA 41, 44, 50, 193
Dolly the sheep 46
dominant genes 43
dopamine 23
double blind tests 37
double decomposition 125
drug testing 37
drugs 32,37
dust in the atmosphere 196

E numbers 87
earth 112–14, 201–3
Earth Summit (Brazil) 67
earthquakes 114, 183–4
earth's crust 116
echo sounding 179
ecosystems 57
efficiency, energy transfer 160–1
eggs
    cooking 80, 88
    reproduction 20
electricity 161, 165–7, 169–70
electrolysis 118, 121
electromagnetic radiation 179–80, 187–8, 196
electrons 72, 149
electrovalent bonding 74
elements 73, 105
embryos 45

emulsions 76–7, 87
endothermic reactions 136–8
energy 62, 136–8, 160–1, 187–8
environment 47, 66–7
enzymes 135
equations 81–2, 177
esters 101
ethanol 100–1
ethene 98
ethics 24, 37, 52, 69
eutrophication 106
evolution 48–50, 51
examinations 6
exothermic reactions 136–8
exploitation 65–6
exponential growth of populations 63
extinction 65–6
extraction of metals 118
eyes 13–14

fermentation 100
fertilisers 105–6
fertility 22
filament lamps 170
fission, nuclear 151–2
flowers 59
focus, eye 13
food 9–10, 61–2, 79–80, 87–8
formulae of atoms 78
fossil fuels 103, 161, 196
fossils 48
fractional distillation 94–5, 103
frequency 176
fuels 100, 103, 161–2, 196
fungi 56

galaxies 208
gamma (γ) radiation 145, 148
Gamow, George 215
gas tests 132
genes 41–4, 44–6, 52
geothermal energy 163
glands 18
global warming 68, 92, 160, 195–6
glucose 61
Gore-Tex 108
grade improvement 7
graphite 139
gravity 206–7
Great Britain (radioactivity map) 146
greenhouse effect 64, 164
group 1 elements (alkali metals) 83–4
group 7 elements (halogens) 85–6
groups (periodic table) 82

Haber process 104–5
habitats 57
haematite (iron ore) 119, 141
half-life of isotopes 150–1
halogens 85–6
health 27–8
heart disease 35–6
heat 19, 138, 156–9
heating 156–7
helium 139, 150
HEP see hydroelectric power
hertz (Hz) unit 176
hexane 98
high density lipoproteins (HDL) 36

homeostasis 17–19
hormones 17–18, 20–2
hosts 60
how science works 8
Hoyle, Fred 216
Hubble, Edwin 215
Human Genome Project 41
hunting 65–6
Huntington's disorder 44
hydroelectric power (HEP) 162
hydrogen 132
hydrogen sulfide 103

identical twins 42
immunity 30
infectious diseases 27
infertility 24
infrared radiation 192–3
inheritance 43–4, 47
insecticides 106
insoluble salts 125
insulation 159
insulin 18
intelligent packaging 88
intensity of radiation 186–7
interference of waves 181–2
iodine 139
ionic bonds 74, 76
ionising radiation 146, 203
ions 73
iris recognition 193
iron 119–21, 120
isotopes 73, 149–50, 150–1

Kevlar (plastic) 108
kingdoms classification 55–6
kwashiorkor (disease) 36

Lamarck 51
lasers 193
latent heat 157
LCA see life cycle assessment
lichens 65
life 50, 55–7, 212
life cycle assessments (LCAs) 140
light 134, 192–3, 210
light-dependent resistors (LDRs) 170
light years 209–10
lime 93
limestone 93–4, 114
limewater 92
limiting factors in photosynthesis 61
lions 57
lithium (Li) 83–4
living together 57–62
longitudinal waves 177
low density lipoproteins (LDLs) 36
lung cancer 33

maggots (rat-tailed) 65
magnesium oxide 74
magnetic fields 165–6, 202–3
manned missions 211
mantle of earth 112
manufacture of metals 117–21
marble 93–4
measles mumps and rubella (MMR)
    vaccination 38
melting points 138–9, 157

memory cells 30
metals 86, 116–21
meteors 204–5
methane 163
microwaves 190–1, 213
Milky Way galaxy 208–9
mixtures 76–7
MMR see measles mumps and rubella
mobile phones 191, 197
molecules 75–6
monthly cycle 21
Moon 204
moths 49–50
mules 55
mutations 29, 47–8, 49–50
mutualism 60

names, compounds 77–8
National Grid 168–9
natural gas 137
natural selection 49–50
near earth objects (NEOs) 205
negative feedback 18
NEOs see near earth objects
nervous system 16–19
neurones 14–15
neutralisation 123
neutrons 72, 149
nicotine 32
nitinol (alloy) 122, 126
nitrates 106
nitrogen 104–6
non-destructive testing 148
non-infectious diseases 27
non-metals 91–2, 117
nuclear fission 151–2
nuclear fuels 162
nuclear reactors 171
nuclei 149–50

obesity 34
oceanic plates 113
oestrogen 20–1
Ohm's Law 169–70
oil (crude) 94–5
optic nerve 13
oral contraceptives 22
orbits 206
ores 117–18
organic compounds of carbon 95–8
oven (microwave) 190
ovulation 20
oxidation 131
oxygen
    covalent bonding 75
    debt 11
    identification of gases 132
    reactions with group 1 elements 84
ozone 64, 194

P-waves 177–8
packaging (food) 87–8
Panthera species 57
paper chromatography 81
paper thickness detectors 147–8
parallax 210
parasites 60
Parkinson's disease 23
particle size 134

Index

passive immunity 30
passive solar heating 164
payback time for insulation 159
penicillin 29
peppered moths 49–50
periodic table 82–6
peripheral nervous system 16
petrol 94
photosynthesis 61, 137
physical properties
   group 1 elements 83–4
   group 7 elements 85
   metals and non-metals 117
pituitary gland 21
placebos 37
planets 204
plant kingdoms 56
plasticisers 115
plastics 100
plate tectonics 112–14, 201–2
polar bears 59
pollination 59
pollution 63–4, 102–3
poly vinyl chloride (PVC) 99
polyatomic ions 74
poly(ethene) 99
polymers 98–101, 108, 115
populations 57, 63
post-it notes 107
potassium (K) 83–4
power 166–7, 169, see also nuclear
   fission
practical skills 8
precautionary principle 197
precipitation 125, 131
predators 59–61
pressure 134
prey 59–61
Proctoctista 56
progesterone 20–1
protection by vaccines 31
protons 72, 149
pyramid of biomass 62

quadrats 58
questions (reading) 7
quicklime 93

radiation
   dangers 146–7
   exposure 153
   heat transfer 158
   intensity 186–7
radio waves 189
radioactivity 145–6, 147–8, 150–2
rat-tailed maggots 65
reaction rates 133–4
reactions, energy transfer 136–8
reactors (nuclear) 171
recessive genes 43
recycling of metals 122–3
red giants (stars) 207
red shift 213
reduction 131
reflection of waves 180, 182
reflex action 16
refraction of waves 181
renewable fuels 162

reproduction 20–2, 41–2, 446
respiration 11, 61
responses 12, 16–17
revision 6
risks
   infertility 24
   radiation 146
   vaccinations 38
rocks 114–16, 116–21, 201
rust 120–1, 133, 135

S-waves 177–8
safety (radiation) 146
salts 123–5
sandstone 114
Sankey energy flow diagram 160
SARs see specific absorption rates
scientific discoveries 107
scientific language 7
scurvy 36
sea floor 202
Search for Extra-Terrestrial Intelligence
   (SETI) 212
seismic waves 177–8
SETI see Search for Extra-Terrestrial
   Intelligence
sex determination 42–3
sexual reproduction 42
Shapley, Harlow 215
shivering 19
skin 26
skin cancer 198
slaking 137
slate 114
slime 99
Smart Memory Alloys (SMAs) 126
smoke detectors 147
smoking 23, 32–3
sodium carbonate 139
sodium chloride 74
sodium (Na) 83–4
solar energy 163–4, 164
solar flares 203–4
solar system 204–5, 211
solder 122
sound 179
space travel 210–11
species, classification 55
specific absorption rates (SARs) 197
specific heat capacity 156
speed of reactions see reactions rates
speed of waves 176
sperm 20
stainless steel 122
stars 207–8
start of life 50
state symbols 82
steel 119, 122
stem cells 46
sterilisation by radioactivity 148
stimuli response 12
storage of radioactive waste 152
studying 6
subduction 113
sudan 1 dye 81
sulfuric acid 64
Sun, solar system 204
sun and skin cancer 198

sunscreen 194
superglue 107
surface area 135
sustainable development 67
sweating 19
symbol equations 81–2
synapses 14–15

technetium 150
Teflon 107
telescopes 211–12
temperature 19, 134–5, 156
testosterone 20
therapeutic cloning 46
thermal decomposition 131
thermistors 170
thermograms 156
thermoplastics 99
tides 163
tigers 57
tissue culture 45
titration 124
tobacco 32
tongue rolling genes 43–4
total internal reflection of waves 182
tracers (radioactive) 148
transformers 168–9
transition elements (metals) 83, 86
transverse waves 176–7
tsunamis 113

ultrasound 179
ultraviolet radiation 194–5
units of electrical energy 167
universe 208–9, 215–16
unmanned missions 211
uranium 150–1

vaccinations 30–1, 38
variable resistors 169
variation 47–8
vasoconstriction 19
vasodilation 19
vertebrates 56
viruses 27
volcanoes 114
voluntary action 16

waste disposal (radioactivity) 152
water
   microwaves 190
   reactions with group 1 elements 84
   reactions with group 7 elements 85
waves 176–82, 188–9
whales 66–7, 69
white blood cells 28
wind pollination 59
wind turbines 163
word equations 81
workings 7

X chromosomes 43
X-rays 195

Y chromosomes 43